ALGEBRA

PATTERNS AND GRAPHS

Walker Maths Essentials: Algebra Patterns and Graphs 5
1st Edition
Charlotte Walker
Victoria Walker

Cover design: Cheryl Smith, Macarn Design
Text design: Cheryl Smith, Macarn Design
Production controller: Siew Han Ong

Any URLs contained in this publication were checked for currency during the production process. Note, however, that the publisher cannot vouch for the ongoing currency of URLs.

Acknowledgements
Cover photo courtesy of Shutterstock.

We wish to thank the Boards of Trustees of Darfield and Riccarton High Schools for allowing us to use materials and ideas developed while teaching. Our thanks also go to all past and present colleagues, especially Kath Wilson, who have generously shared their experience and ideas.

For product information and technology assistance,
in Australia call **1300 790 853**;
in New Zealand call **0800 449 725**

For permission to use material from this text or product, please email **aust.permissions@cengage.com**

National Library of New Zealand Cataloguing-in-Publication Data
A catalogue record for this book is available from the National Library of New Zealand.

978 0 17 045146 8

Cengage Learning Australia
Level 7, 80 Dorcas Street
South Melbourne, Victoria Australia 3205

Cengage Learning New Zealand
Unit 4B Rosedale Office Park
331 Rosedale Road, Albany, North Shore 0632, NZ

For learning solutions, visit **cengage.co.nz**

Printed in China by 1010 Printing International Limited.
2 3 4 5 6 7 25

CONTENTS

Glossary

Make your own glossary of key terms:

Term	Definition	Picture/Example
Expression		
Variable		
Term		
Sequence		
Origin		
Coordinate		
Horizontal		
Vertical		
Linear		
Non-linear		
Quadratic		
Parabola		
Discrete data		
Continuous data		

ISBN: 9780170451468

Revision: substitution

- When substituting, you replace **variables** (letters) with **numbers**.
- Sometimes you will need to apply BEDMAS.

Examples: If $n = 4$, find the values of A.

1

$A = n + 7$
$= \mathbf{4} + 7$
$A = 11$

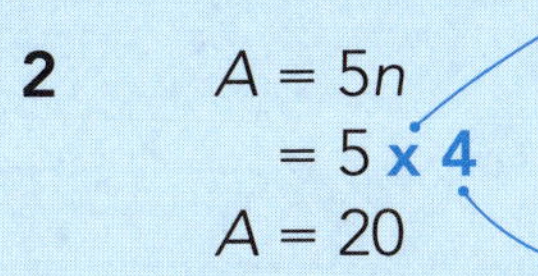

2

$A = 5n$
$= 5 \times \mathbf{4}$
$A = 20$

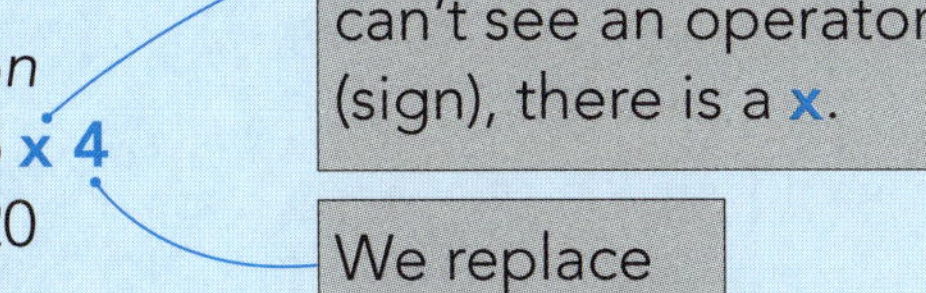

3

$A = 6n - 13$
$= 6 \times \mathbf{4} - 13$
$= 24 - 13$
$= 11$

4

$A = -3n - 10$
$= -3 \times \mathbf{4} - 10$
$= -12 - 10$
$= -22$

Complete the following table.

Formula	$n = 3$	$n = 6$	$n = -2$
$A = n + 4$	$A =$ ______ $=$ ______	$A =$ ______ $=$ ______	$A =$ ______ $=$ ______
$A = 6n$	$A =$ ______ $=$ ______	$A =$ ______ $=$ ______	$A =$ ______ $=$ ______
$A = 2n + 3$	$A =$ ______ $=$ ______	$A =$ ______ $=$ ______	$A =$ ______ $=$ ______
$A = 3n - 5$	$A =$ ______ $=$ ______	$A =$ ______ $=$ ______	$A =$ ______ $=$ ______
$A = -2n - 4$	$A =$ ______ $=$ ______	$A =$ ______ $=$ ______	$A =$ ______ $=$ ______

ISBN: 9780170451468

Patterns

- Patterns can be seen all around us, often in art and cultural decorations.

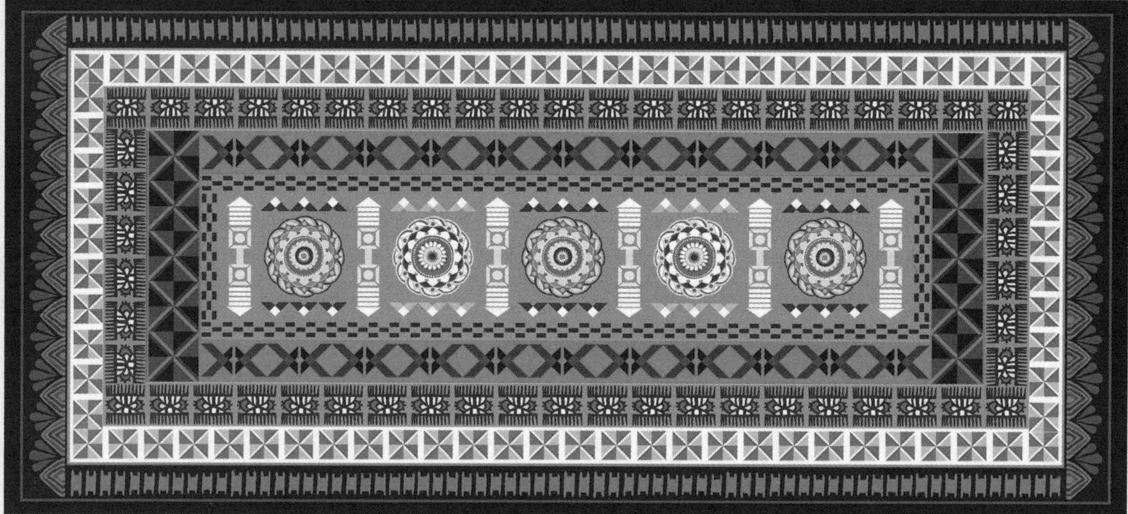

Linear patterns

- '**Linear**' means the pattern increases or decreases by the same amount each term, e.g. 2, 4, 6, 8, 10, … or 100, 97, 94, 91, 88, …
- When plotted on a graph, a linear pattern forms a **straight line**.
- A '**term**' is one element of a pattern. For example, for the pattern 2, 4, 6, 8, 10, …, the first term is 2, the second term is 4, etc.
- Patterns can be expressed using different methods:
  - — in a series of pictures
  - — as a list
  - — as an equation
  - — in a table
  - — using words
  - — on a graph.
- You will need to convert between any of these methods.

ISBN: 9780170451468

Patterns into tables

- Tables are a very useful way of showing patterns.
- We can use tables:
  1 to show us the pattern, and
  2 to predict, for instance, the number of popsicle sticks needed for bigger shapes **without having to draw them**.

Example:

Shape 1 Shape 2 Shape 3

Shape number	Number of popsicle sticks
1	6
2	9
3	12
4	**15**
5	**18**
6	**21**

+ 3 + 3 + 3 + 3 + 3

To find the number in the next term, we need to **add 3**. You can see this in the shapes: **3 more** popsicle sticks were added each time.

Because we added 3 for the first row, we **keep adding 3** to complete the table — **without** having to draw the pictures.

Complete the tables for the shapes that have been drawn (the white boxes). Then, without drawing the next shapes, fill in the grey boxes.

1

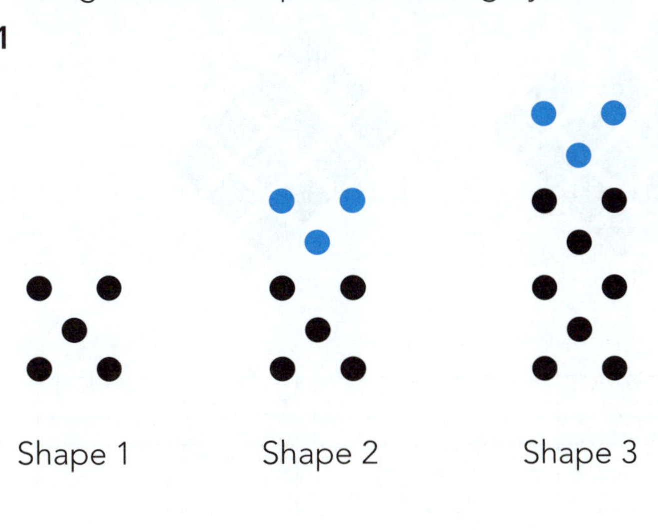

Shape 1 Shape 2 Shape 3

Shape number	Number of dots
1	5
2	
3	
4	
5	
6	

ISBN: 9780170451468

2

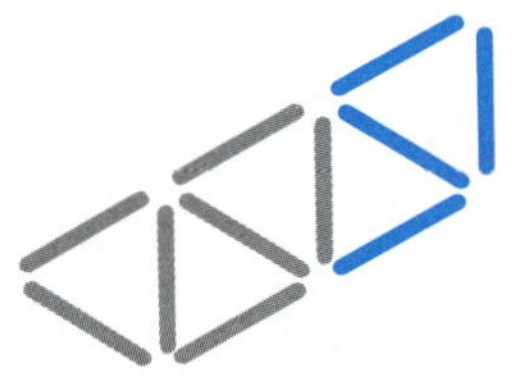
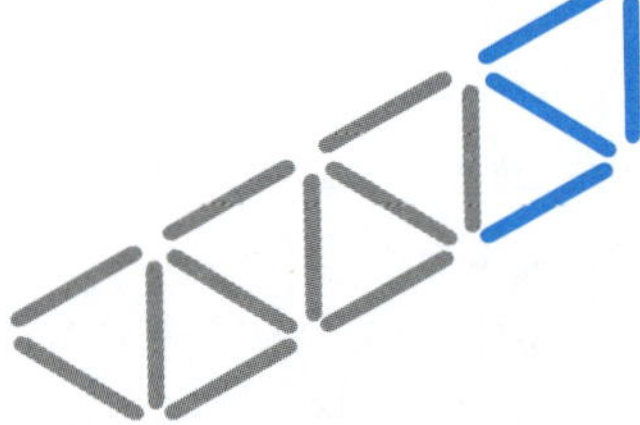

Shape number	1	2	3	4	5	6
Number of popsicle sticks						

3

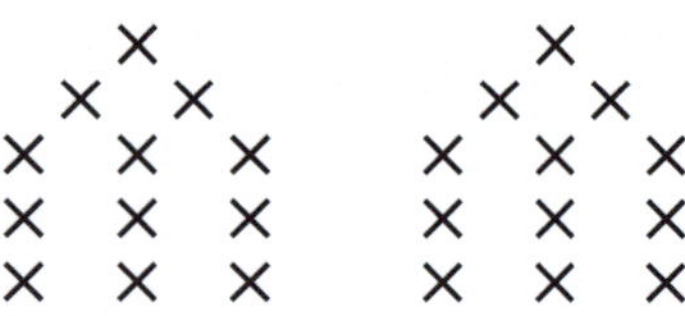
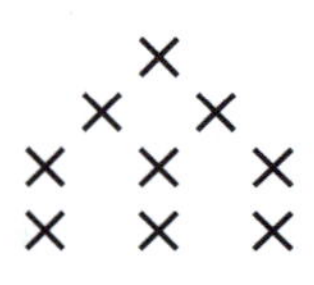

Shape number	Number of crosses
1	15
2	
3	
4	
5	
6	

4

Shape number	1	2	3	4	5	6
Number of diamonds						

ISBN: 9780170451468

Finding term 'zero'

- Not all patterns involve **shapes**. For example, the odd numbers between 10 and 20 form a pattern: 11, 13, 15, 17, 19.
- So from now on we will use '**term**' instead of 'shape number'. E.g. term 1 = 11, term 4 = 17.
- It will also be very useful to find the **value of term 0**.

Example:

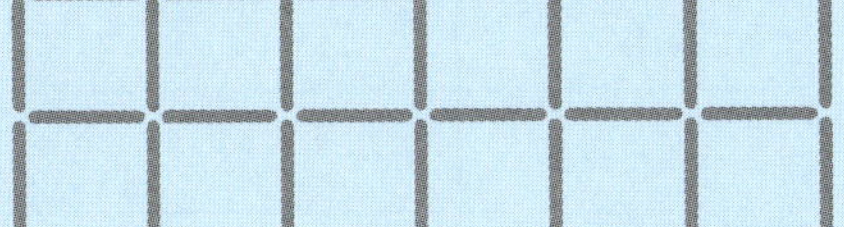

Often there is no practical meaning for the term 0, but it's useful for finding the equation.

Term number	Number of popsicle sticks
0	**37**
1	32
2	27
3	22
4	17
5	12

+ 5, – 5, – 5, – 5, – 5

To find the bigger terms we **subtract 5**. So to find term zero we need to **add 5** to the first term.
32 **+ 5** =37

Complete the tables, including the value for term zero.

1

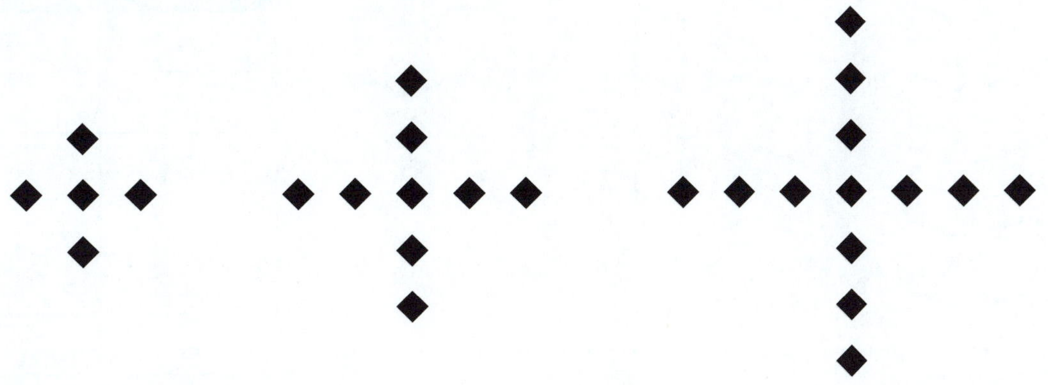

Term number	0	1	2	3	4	5	6
Number of diamonds		5					

ISBN: 9780170451468

2

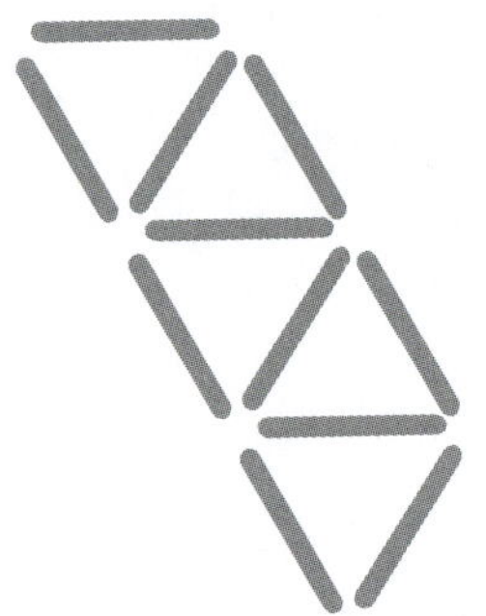

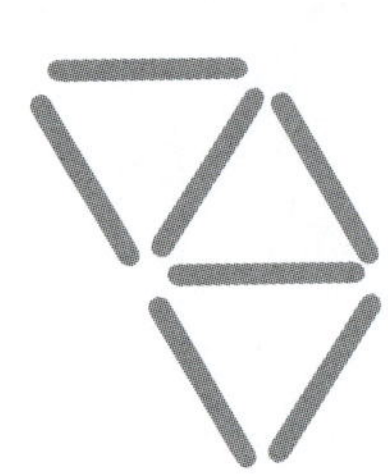

Term number	Number of popsicle sticks
0	
1	
2	9
3	
4	
5	

3

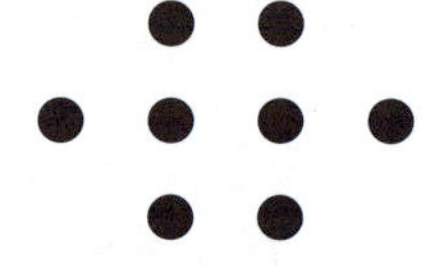
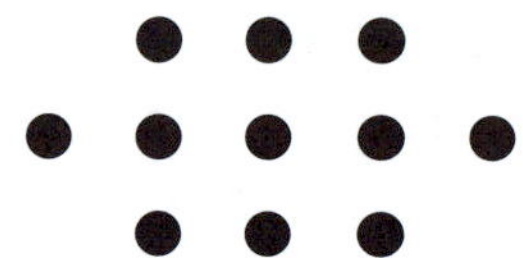
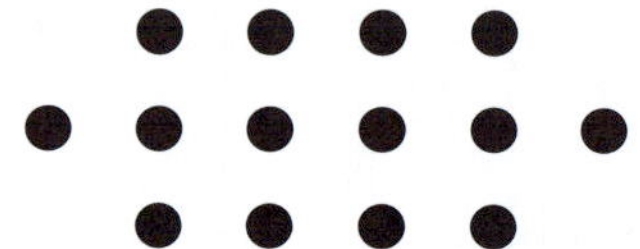

Term number	0	1	2	3	4	5	6
Number of dots							

4

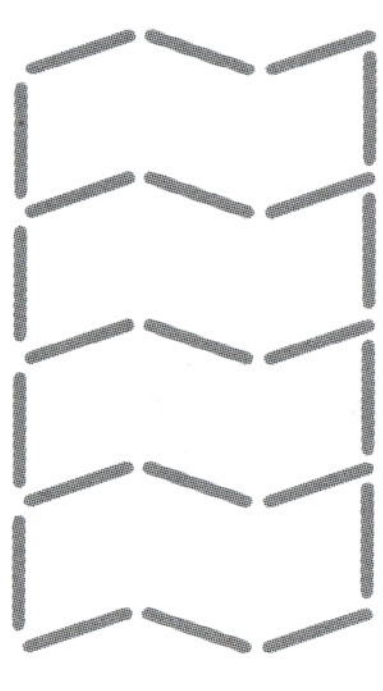
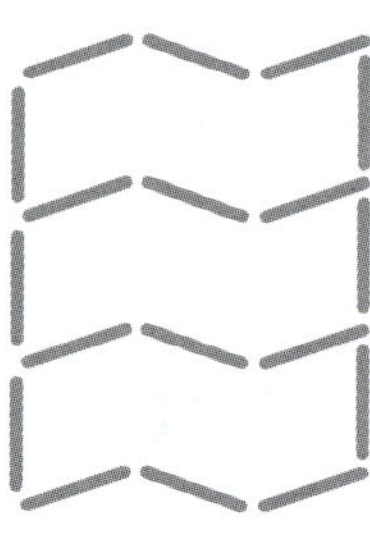
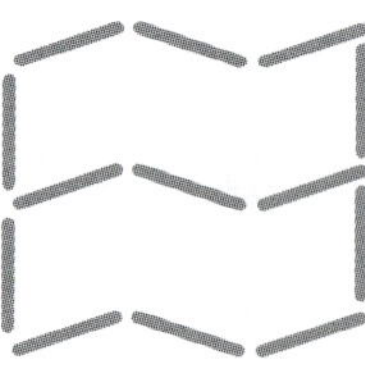

Term number	Number of popsicle sticks
0	
1	
2	
3	
4	
5	

ISBN: 9780170451468

Finding the rule from a pattern

Patterns and tables

Think about the answers to the times tables:

Table	0	1	2	3	4	5	6	Change at each step of the table
x 3	0	3	6	9	12	15	18	+ 3
x 4	0	4	8	12	16	20	24	+ 4
x 5	0	5	10	15	20	25	30	+ 5

Now substitute the values of *n* into these expressions:

Value of n	0	1	2	3	4	5	6	Change at each step in the value of n
$4n$	0	4	8	12	16	20	24	+ 4
$4n + 3$	3	7	11	15	19	23	27	+ 4
$4n - 1$	–1	3	7	11	15	19	23	+ 4

So, if a pattern goes up in fours, then the equation will contain a $4n$.

Decreasing patterns

Table	0	1	2	3	4	5	6	Change at each step of the table
x (–3)	0	–3	–6	–9	–12	–15	–18	– 3
x (–4)	0	–4	–8	–12	–16	–20	–24	– 4
x (–5)	0	–5	–10	–15	–20	–25	–30	– 5

Now substitute the values of *n* into these expressions:

Value of n	0	1	2	3	4	5	6	Change at each step in the value of n
$-5n$	0	–5	–10	–15	–20	–25	–30	– 5
$-5n + 2$	2	–3	–8	–13	–18	–23	–28	– 5
$-5n - 4$	–4	–9	–14	–19	–24	–29	–34	– 5

So, if a pattern goes down in fives, then the equation will contain a $-5n$.

- It is useful to have a **mathematical rule** for patterns so we can work out, for instance, the value of the 100th term **without** having to draw pictures or create a large table.
- Mathematical rules are easiest if they are written using **letters** (which we call **variables**) rather than words.

Example:

Term number (n)	Number of popsicle sticks (P)
0	3
1	7
2	11
3	15
4	19
5	23

– 4, + 4, + 4, + 4, + 4

We could draw term number 0: it would be just 3 popsicle sticks.

Remember, do the **opposite** (– 4) to find term 0.

In words: I started with 7 popsicle sticks and then I **added** 4 each time.

Find the rule: Number of **P**opsicle sticks = 4 x term **n**umber + 3

Tidy it up: $P = 4n + 3$

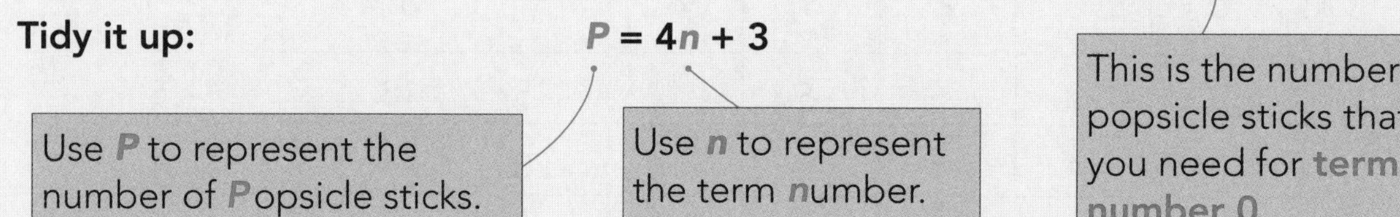

Use **P** to represent the number of **P**opsicle sticks.

Use **n** to represent the term **n**umber.

This is the number of popsicle sticks that you need for **term number 0**.

Word rule: The number of **P**opsicle sticks is calculated by multiplying the term **n**umber by **4** and then **adding 3**.

Use the rule: To find the number of popsicle sticks needed for the 50th shape:

$n = 50$ so $P = 4n + 3$

$= 4 \times 50 + 3$

$= 203$

We have found the number of popsicle sticks needed for the 50th shape **without** having to draw all the shapes or create a huge table.

 ISBN: 9780170451468

Example of a decreasing pattern:

Term number (n)	Number of dots (D)
0	22
1	19
2	16
3	13
4	10

+ 3, – 3, – 3, – 3

We could draw term number 0: it would have 22 dots:

Remember, do the **opposite** (+ 3) to find term 0.

In words: I started with 19 popsicle sticks and then I **subtracted** 3 each time.

Find the rule: Number of **D**ots = – 3 x term **n**umber + 22

Notice that for reducing patterns, we need to multiply the term number by a **negative** number.

This is the number of popsicle sticks that you need for **term number 0**.

Tidy it up: $D = -3n + 22$

Word rule: The number of **D**ots is calculated by multiplying the term **n**umber by **–3** and then **adding 22**.

Use the rule: To find the number of dots needed for the 7th shape:

$n = 7$ so $D = -3n + 22$

$= -3 \times 7 + 22$

$= 1$

Again, we have found the number of dots needed for the 7th shape **without** having to draw all the shapes or create a larger table.

ISBN: 9780170451468

Complete the tables and fill in the gaps.

1

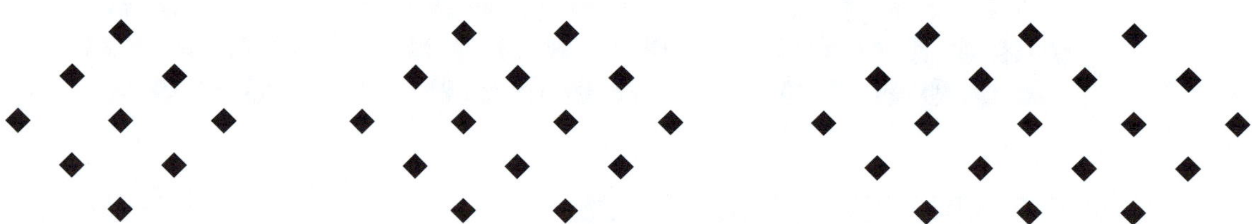

Complete the table:

Term number (n)	Number of diamonds (D)
0	
1	9
2	
3	
4	
5	
6	

In words:
I started with ________ diamonds and then I added/subtracted ________ each time.

Find the mathematical rule:

Number of diamonds = ________ x term number ________

Tidy it up: $D =$ ________n ________

Word rule:
The number of diamonds is calculated by multiplying the term number by ________ and then adding/subtracting ________.

Use the rule:
How many diamonds would be needed for the 50th shape?

$D =$ ________________

$=$ ________________

 ISBN: 9780170451468

2

Complete the table:

Term number (n)	0	1	2	3	4	5	6
Number of squares (S)							

In words:
I started with ______ squares and then I added/subtracted ______ each time.

Find the mathematical rule:

Number of squares = ______ x term number ______

Tidy it up: $S =$ ______

Word rule:
The number of squares is calculated by multiplying the term number by ______ and then adding/subtracting ______.

Use the rule:
How many squares would be needed for the 10th shape?

$S =$ ______

$=$ ______

ISBN: 9780170451468

3

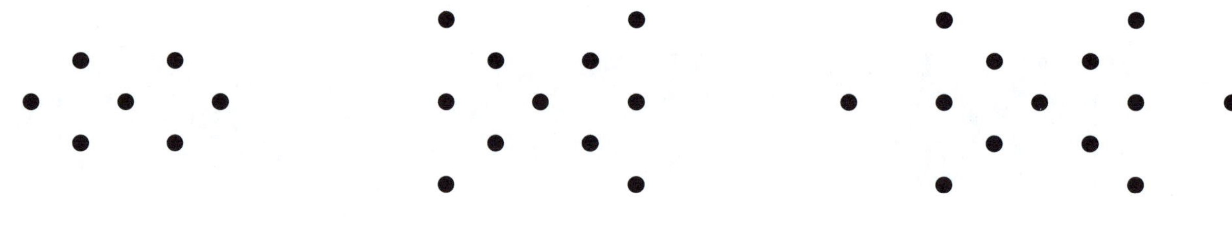

Complete the table:

Term number (n)	Number of dots (D)
0	
1	
2	
3	
4	
5	
6	

In words:
I started with ________ dots and then I added/subtracted ________ each time.

Find the mathematical rule:

Number of dots = ________ x term number ________

Tidy it up: D = ____________________

Word rule:
The number of dots is calculated by multiplying the term number by ________ and then adding/subtracting ________.

Use the rule:
How many dots would be needed for the 40th shape?

D = ____________________

= ____________________

 ISBN: 9780170451468

4

Complete the table:

Term number (n)	Number of popsicle sticks (P)
0	
1	
2	
3	
4	
5	
6	

In words:
I started with ________ popsicle sticks and then I added/subtracted ________ each time.

Find the mathematical rule:

Number of popsicle sticks = ________ x term number ________

Tidy it up: $P =$ ____________________

Word rule:
The number of popsicle sticks is calculated by multiplying the term number by ________ and then adding/subtracting ________.

Use the rule:
How many popsicle sticks would be needed for the 20th shape?

$P =$ ____________________

$=$ ____________________

ISBN: 9780170451468

5

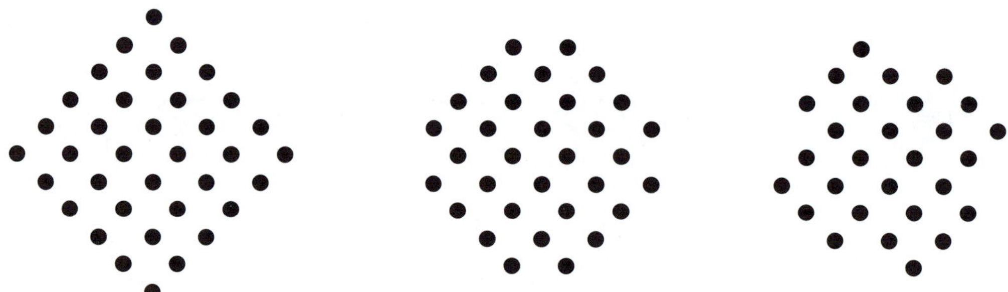

Complete the table:

Term number (n)	Number of dots (D)
0	
1	
2	
3	
4	
5	
6	

In words:
I started with ________ dots and then I added/subtracted ________ each time.

Find the mathematical rule:

Number of dots = ________ x term number ________

Tidy it up: D = ____________________

Word rule:
The number of dots is calculated by multiplying the term number by ________ and then adding/subtracting ________.

Use the rule:
How many dots would be needed for the 8th shape?

D = ____________________

= ____________________

 ISBN: 9780170451468

Finding the rule from a table

- As we said earlier, not all patterns involve shapes: patterns are sometimes **described**.
- In these cases we need to use the description to draw up a table.

Examples:

1 Complete the table and find the rule for a pattern which starts at 8 and increases in threes.

Draw up a table:

Term number (n)	Value of term (T)
0	**5**
1	8
2	11
3	14
4	17
5	20

– 3
+ 3
+ 3
+ 3
+ 3

Find the rule:

The pattern **increases** by **3** each time.

$T = 3 \times n + 5$

This is the value of term zero.

Tidy it up:

$\mathbf{T = 3n + 5}$

Use the rule:
Find the value of the 100th term.

$\mathbf{T = 3n + 5}$
$= 3 \times 100 + 5$
$= 305$

2 Complete the table and find the rule for a pattern which starts at 92 and decreases in fives.

Draw up a table:

Term number (n)	0	1	2	3	4	5	6
Value of term (T)	**97**	92	87	82	77	72	67

+ 5 – 5 – 5 – 5 – 5 – 5

The pattern **decreases** by **5** each time.

Find the rule: $T = -5 \times n + 97$

Tidy it up: $\mathbf{T = -5n + 97}$

Use the rule:
Find the value of the 15th term.

$\mathbf{T = -5n + 97}$
$= -5 \times 15 + 97$
$= 22$

ISBN: 9780170451468

Complete the tables and find the rules for the following.

1 Increasing odd numbers that are larger than 20.

Term number (n)	Value of term (T)
0	
1	21
2	
3	
4	

Rule:

$T =$ ______ $n +$ ______

______

2 Start at 19, and add 4 each time.

Term number (n)	0	1	2	3	4	5
Value of term (T)						

______

Rule:

$T =$ ______ n ___ ______

3 Increasing multiples of 7 that are bigger than 1.

Term number (n)	0	1	2	3	4	5
Value of term (T)						

Rule:

$T =$ ______ n ___ ______

4 Increasing multiples of 12 that are bigger than 13.

Term number (n)	Value of term (T)
0	
1	
2	
3	
4	

Rule:

$T =$ ______________

5 Decreasing even numbers that are smaller than 54.

Term number (n)	0	1	2	3	4	5
Value of term (T)						

Rule:

$T =$ ______________

ISBN: 9780170451468

6 Start at 79 and go down 3 each time.

Rule:

Term number (n)	Value of term (T)
0	
1	
2	
3	
4	

$T =$ ______

7 Start at –11 and go up in 7s.

Rule:

Term number (n)	1	2	3	4	5
Value of term (T)					

$T =$ ______

8 Start at 101, then subtract 3 each time.

Rule:

Term number (n)	1	2	3	4	5
Value of term (T)					

$T =$ ______

9 Decreasing multiples of 9, starting at 27.

Rule:

Term number (n)	Value of term (T)
1	
2	
3	
4	

$T =$ ______

10 Decreasing odd numbers, starting at –13.

Rule:

Term number (n)	1	2	3	4	5
Value of term (T)					

$T =$ ______

ISBN: 9780170451468

Finding the rule from a list

- You can find a rule from a table, but finding a rule from just a list saves space and time.

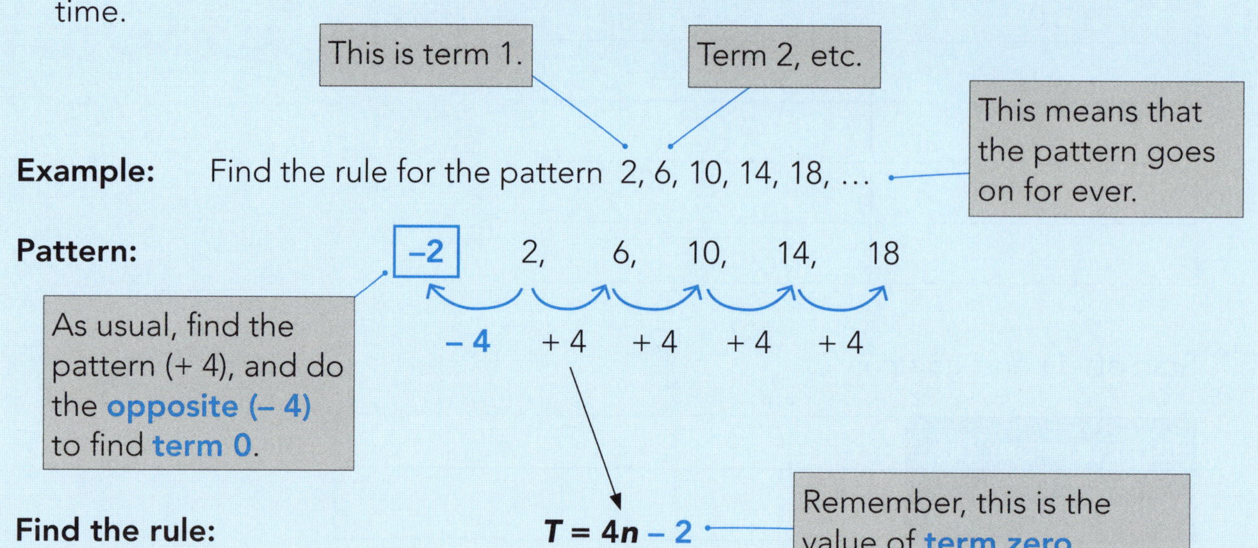

Words: Any value in the pattern can be calculated by multiplying the term number by **4** and then subtracting **2**.

Pick the correct rule from the options below.

1

7, 10, 13, 16, 19, …	
$T = 3n + 3$	$T = -3n + 4$
$T = 3n + 4$	$T = 3n - 4$

2

19, 17, 15, 13, 11, …	
$T = -2n + 19$	$T = -2n + 21$
$T = 21n - 2$	$T = 2n - 19$

3

1, 6, 11, 16, 21, …	
$T = 5n + 1$	$T = -5n - 4$
$T = 5n - 5$	$T = 5n - 4$

4

24, 20, 16, 12, 8, …	
$T = -4n + 28$	$T = 4n - 28$
$T = 28n - 4$	$T = -4n + 24$

5

100, 89, 78, 67, 56, …	
$T = -n + 111$	$T = -11n + 111$
$T = 111n - 11$	$T = -111n - 11$

6

0, 15, 30, 45, 60, …	
$T = 15n + 15$	$T = 15n - 15$
$T = -15n - 15$	$T = -15n + 15$

7

–19, –15, –11, –7, –3, …	
$T = 4n - 23$	$T = -4n - 23$
$T = 23n - 4$	$T = -23n - 4$

8

–3, 3, 9, 12, 15, …	
$T = -6n - 3$	$T = 6n - 9$
$T = -3n - 6$	$T = -6n - 6$

ISBN: 9780170451468

Complete the statements and find the rule for each of these.

9 8, 11, 14, 17, ...

___ ___ ___ ___

Rule:

$T =$ ____ n ____

I started with ____ and then I added/subtracted ____ each time.
Term zero will be ____.

10 1, 4, 7, 10, 13, ...

Rule:

$T =$ ____ n ____

I started with ____ and then I added/subtracted ____ each time.
Term zero will be ____.

11 23, 19, 15, 11, 7, ...

Rule:

$T =$ ____ n __ ____

I started with ____ and then I added/subtracted ____ each time.
Term zero will be ____.

12 –12, –1, 10, 21, 32, ...

Rule:

$T =$ ________

I started with ____ and then I added/subtracted ____ each time.
Term zero will be ____.

13 126, 87, 48, 9, –30, ...

Rule:

$T =$ ________

I started with ____ and then I added/subtracted ____ each time.
Term zero will be ____.

14 –17, 28, 73, 118, 163, ...

Rule:

$T =$ ________

15 –64, –46, –28, –10, 8, ...

Rule:

$T =$ ________

16 –99, 0, 99, 198, 297, ...

Rule:

$T =$ ________

17 0.5, 4.25, 8, 11.75, 15.5, ...

Rule:

$T =$ ________

ISBN: 9780170451468

Finding a term from a rule

- We can use the rule to find the value of any term in the sequence.
- This is much faster than writing a list.

Examples:

1 Find the 70th term for the sequence that has the rule $T = 5n - 3$.

$$T = 5n + 3$$
$$= 5 \times \mathbf{70} + 3$$
$$= 353$$

Substitute **70** for ***n***.

2 Find the 30th term for the sequence that has the rule $T = -3n + 19$.

$$T = -3n + 19$$
$$= -3 \times \mathbf{30} + 19$$
$$= -71$$

Use the rules to calculate the terms.

1 40th term $T = 8n + 15$
= ____________
= ________

2 50th term $T = 4n - 23$
= ____________
= ________

3 20th term $T = -2n + 107$
= ____________
= ________

4 80th term $T = -7n - 16$
= ____________
= ________

5 9th term $T = 16n - 18$
= ____________
= ________

6 19th term $T = -3n - 60$
= ____________
= ________

7 100th term $T = -14n + 70$
= ____________
= ________

8 24th term $T = 17n - 15$
= ____________
= ________

9 14th term $T = -n - 1$
= ____________
= ________

10 7th term $T = -2.5n + 8.25$
= ____________
= ________

ISBN: 9780170451468

Finding the sequence from the rule

- We can use the rule to find the sequence.

Examples:

1 Rule: $T = 4n + 3$

Use substitution, along with the pattern, to find the terms:

Term number (n)	1	2	3	4	5
Calculations	$T = 4 \times 1 + 3$				$T = 4 \times 5 + 3$
Value of term (T)	7	11	15	19	23

+ 4 + 4 + 4 + 4

1 What is the first term? $T = 4 \times 1 + 3 = 7$

2 What does the sequence go up or down by? + 4

3 Check term 5. $T = 4 \times 5 + 3 = 23$ ✓

Tidy it up: 7, 11, 15, 19, 23

2 Rule: $T = -3n - 5$

Term number (n)	Calculations	Value of term (T)
1	$-3 \times 1 - 5$	−8
2		−11
3		−14
4		−17
5	$-3 \times 5 - 5$	−20

− 3 − 3 − 3 − 3

1 What is the first term? $T = -3 \times 1 - 5 = -8$

2 What does the sequence go up or down by? − 3

3 Check term 5. $T = -3 \times 5 - 5 = -20$ ✓

Tidy it up: −8, −11, −14, −17, −20

Find the first five terms of these sequences using the rule.

1 $T = 4n + 7$

Term number (n)	1	2	3	4	5
Calculations	$4 \times 1 + 7$				
Value of term (T)					

Sequence: ______________________________

ISBN: 9780170451468

2 $T = -2n - 8$

Term number (n)	1	2	3	4	5
Calculations					
Value (V)					

Sequence: ______________________________

3 $T = 8n - 11$

Term number (n)	Calculations	Value of term (T)
1	8 x ____ – 11	
2		
3		
4		
5		

Sequence: ______________________________

Write the first five terms for these sequences.

4 $T = -6n + 21$

Sequence: ______________________________

5 $T = n - 1$

Sequence: ______________________________

6 $T = -0.5n + 2$

Sequence: ______________________________

7 $T = 15n - 23$

Sequence: ______________________________

8 $T = 89 - 6n$

Sequence: ______________________________

ISBN: 9780170451468

Applications

- There are many practical situations where it's useful to find the rule.
- From now on, we will use the term **sequence**, rather than pattern.
- A sequence is a list of numbers or objects **in a particular order**.
- Sometime sequences occur in everyday situations.

Example: Emily has \$165 in her bank account. She begins a part-time job, so she is able to deposit \$20 at the end of each week. Find the rule for calculating the total (T) in her account after she has worked for n weeks.

Write a list for the sequence: \$185, \$205, \$225, \$245, ...

Rule: $T = 20n + 165$

Create a list which starts at $n = 1$ and then find the rule for each of these descriptions.

1 **a** Noel has \$50 in his bank account. He begins a part-time job, so he is able to deposit \$25 at the end of each week. Find the rule for calculating the total (T) in his account after n weeks.

$T =$ ____________

____________, ____________, ____________, ...

b Use the rule to calculate the amount he will have in his bank account after a year (52 weeks).

____

____

2 **a** Hiring a mountain bike costs \$25 plus \$14 per hour. Find the rule for calculating the total cost of hiring a bike for n hours.

$T =$ ____________

____________, ____________, ____________, ...

b Use the rule to calculate how much it will cost to hire a bike for 11 hours.

____

____

3 **a** Lee is building a fence which is made up of posts with wire netting stretched between them. Each post needs 7 staples for fastening the netting. He has bought a pack of 500 staples. Find the rule for calculating the number of staples remaining after he has fastened netting to n posts.

$T =$ ____________

____________, ____________, ____________, ...

b Use the rule to calculate how many staples will remain after he has used 34 posts.

____

____

ISBN: 9780170451468

4 **a** Matiu has $40 to spend at the school funfare. Entry costs $2, and each activity costs $4. Find the rule for calculating how much money he has left after he has done n activities.

$T =$ ______

______, ______, ______, ...

b Use the rule to calculate the amount he will have left after he has done 8 activities.

______

______

5 **a** The school sports coordinator has 162 netball dresses. Each team has 9 members. Find the rule for calculating how many dresses remain after she has supplied dresses to n teams.

$T =$ ______

______, ______, ______, ...

b Use the rule to calculate how many dresses remain after she has supplied them to 11 teams.

______

______

6 **a** The sports coordinator would also like to buy 2 football goals at $280 each and n footballs at $35 each. Find the rule for calculating how much it will cost for 2 goals and n footballs.

$T =$ ______

______, ______, ______, ...

b Use the rule to calculate the cost of 2 goals and 17 footballs.

______

______

7 **a** A local club has donated $900 to be used to promote croquet as a school sport. A set of croquet balls and the equipment needed to set up an area for playing cost $335. Mallets cost $48 each. Find the rule for calculating how much money remains after they have bought the croquet balls and equipment plus n mallets.

$T =$ ______

______, ______, ______, ...

b Use the rule to calculate how much money remains after they have bought the croquet balls and equipment plus 8 mallets.

______

______

c Challenge: Show how the equation could be used to calculate the maximum number of mallets they could buy.

______

______

______

______

 ISBN: 9780170451468

Graphs

Coordinate revision

- A positive **x** coordinate tells you how far to move to the **right**.
- A positive **y** coordinate tells you how far to move **up**.
- Coordinates are written in brackets, and in **alphabetical order**: **(x, y)**.
- We use the word 'ax**i**s' for one axis, and the word 'ax**e**s' for more than one.

Write down the coordinates of the lettered points shown below.
The first one is done for you.

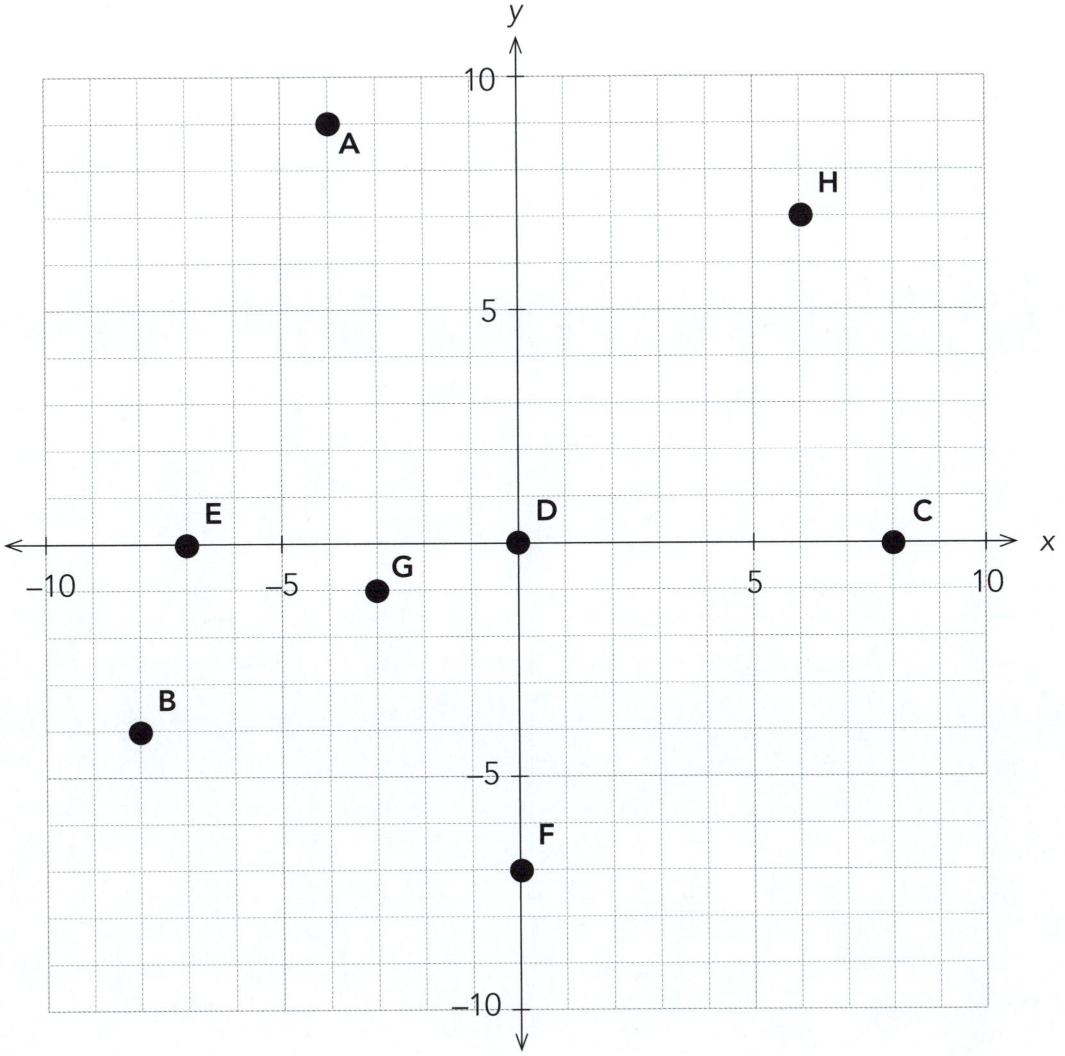

A	(−4 , 9)	**B**	(______, ______)
C	(______, ______)	**D**	(______, ______)
E	(______, ______)	**F**	(______, ______)
G	(______, ______)	**H**	(______, ______)

ISBN: 9780170451468

Plot these points on the axis:

I	(–8, 9)		**J**	(0, -6)
K	(7, 0)		**L**	(–2, -4)
M	(6, –9)		**N**	(–3, 10)
O	(–8, 0)		**P**	(0, –1)

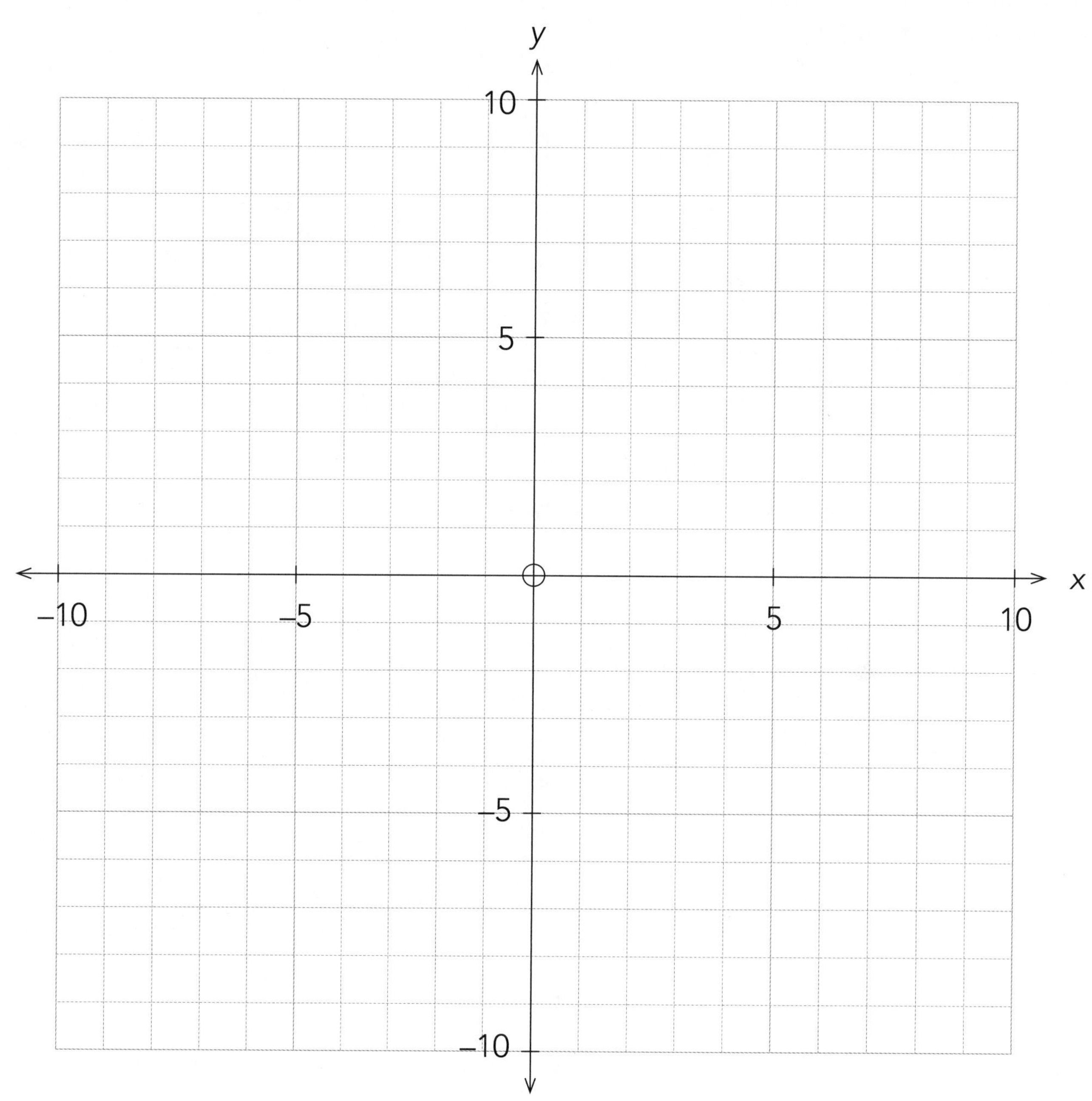

 ISBN: 9780170451468

Plotting discrete patterns

- **Discrete** data is obtained by **counting**: it consists of whole numbers, **no fractions or decimals**.
- **Discrete data** is plotted on graphs as **points**, and **not connected** by lines.
- When we plot patterns on a graph, we use the letters x and y rather than n and T.
- While calculating the term 0 is useful for finding the equation it often has no real meaning, so do not plot it.

Examples:

1 Consider the number of squares.

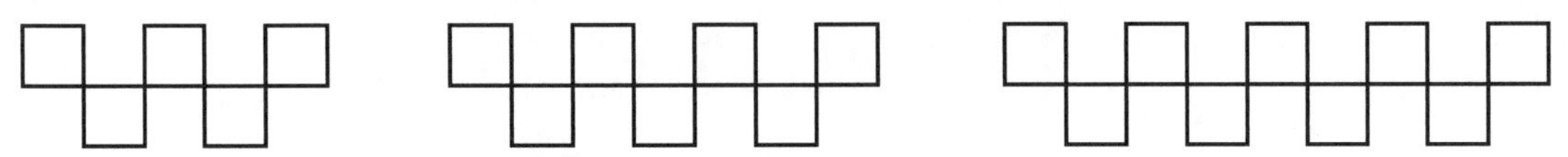

Steps:

1 Put the pattern in a **table**.

For the term number, we use x instead of n.

Term number (x)	Value of the term (y)
1	5
2	7
3	9
4	**11**
5	**13**

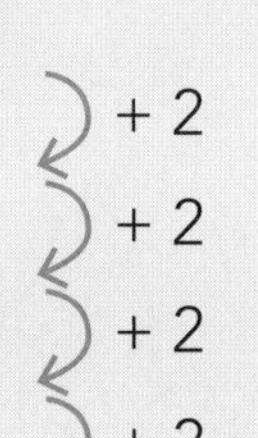

For the value of the term, we use y instead of T.

We use the word **equation** rather than rule.

2 Find the **equation**: $y = \underline{2} \times \text{term number} + \underline{3}$

Tidy it up: $y = 2x + 3$

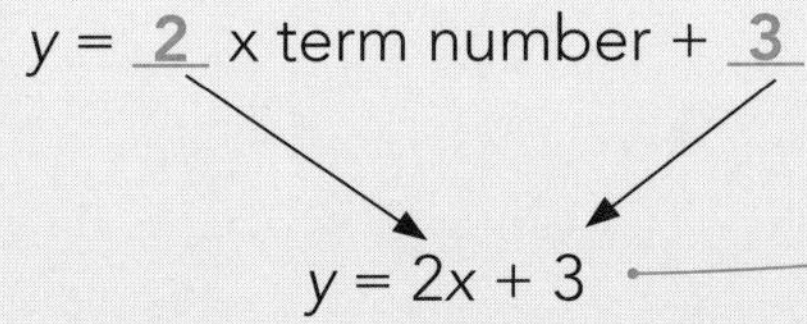

We use the letters x and y in equations.

3 Plot the points on a **graph**:

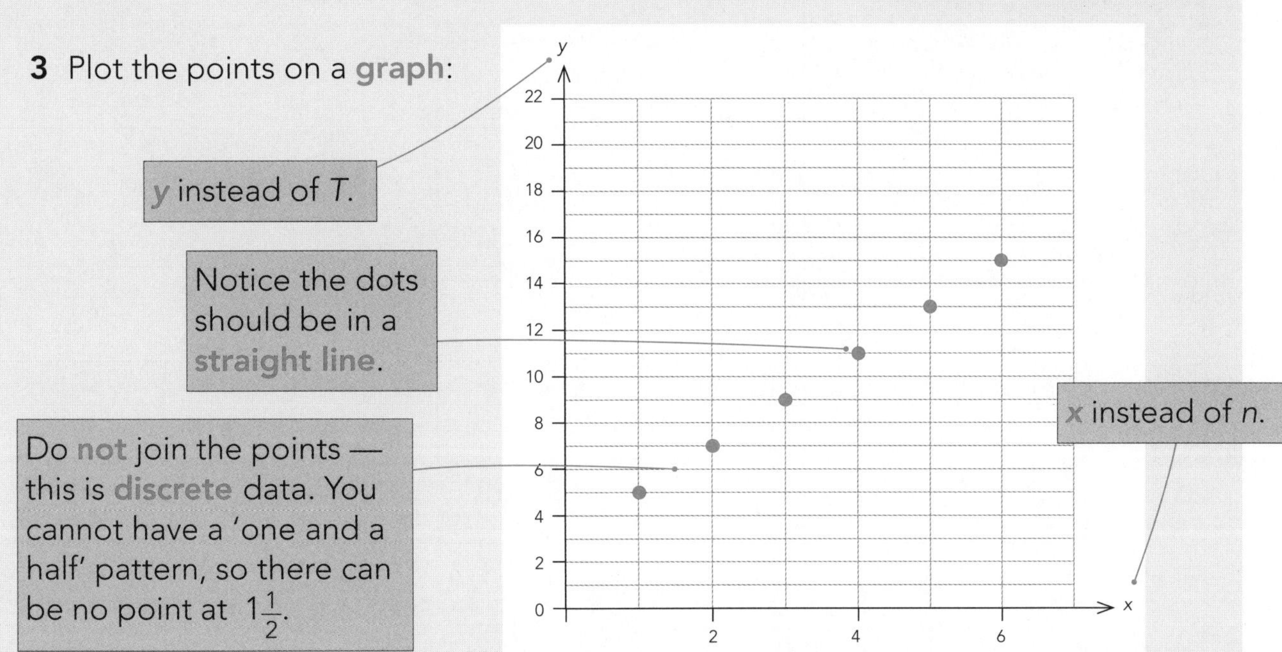

2 Consider the number of popsicle sticks.

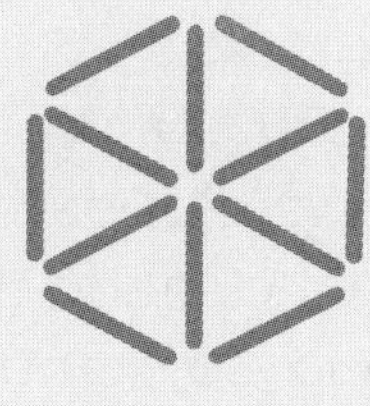

Steps:

1 Put the pattern in a **table**.

Term number (x)	Value of the term (y)
1	24
2	18
3	12
4	6
5	0

−6 −6 −6 −6

2 Find the **equation**: $y =$ <u>−6</u> x term number + <u>30</u>

Tidy it up: $y = -6x + 30$

3 Plot the points on a **graph**:

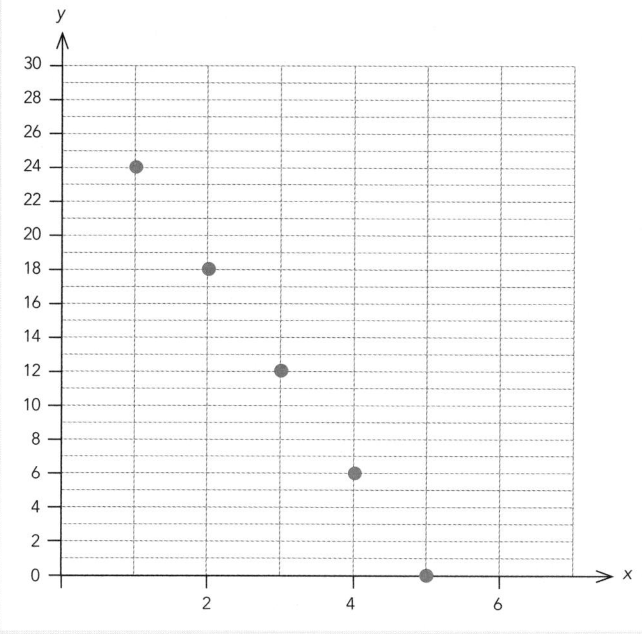

 ISBN: 9780170451468

Complete each table, find the rule, and plot the points on a graph.

1 Consider the number of popsicle sticks.

Term number (x)	Value of the term (y)
1	6
2	
3	
4	
5	

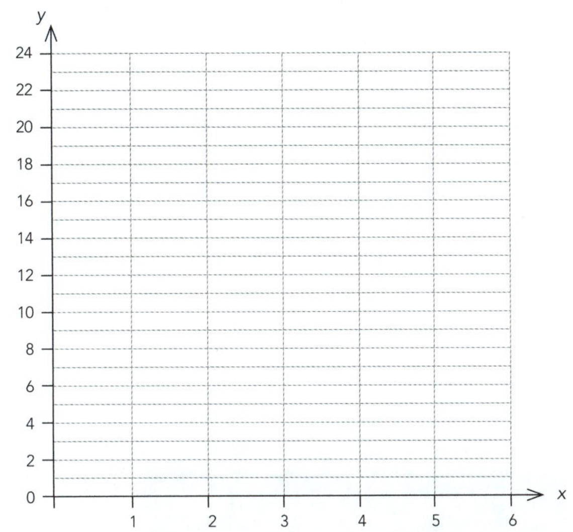

$y =$ ______ x term number + ______

Tidy it up: $y =$ ______x + ______

2 Consider the number of dots.

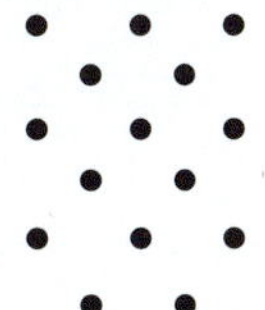
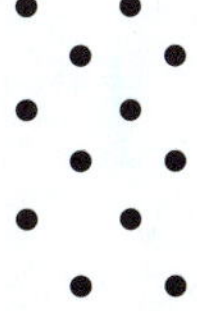

Term number (x)	Value of the term (y)
1	
2	
3	
4	
5	

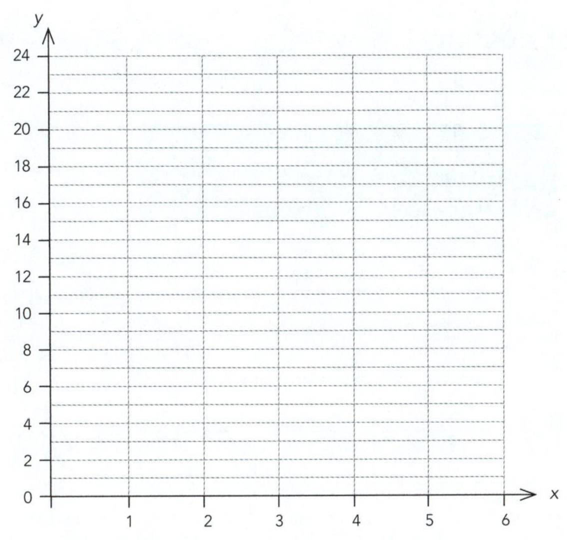

$y =$ ______ x term number + ______

Tidy it up: $y =$ ______x + ______

ISBN: 9780170451468

Complete the table, plot the points and find the rule.

3

Term number (x)	Value of the term (y)
1	6
2	11
3	
4	
5	

Equation: $y =$ ______ $x +$ ______

Sometimes you will get points that don't fit on the axes you are given. If this happens, just leave the point out.

4

Term number (x)	Value of the term (y)
1	20
2	
3	14
4	
5	

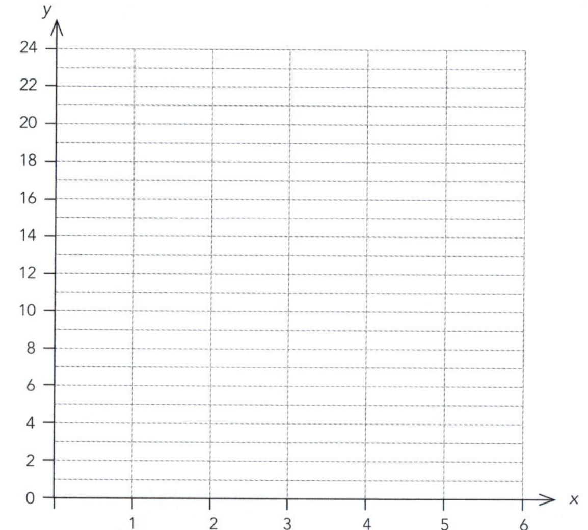

Equation: $y =$ ______ $x +$ ______

5

Term number (x)	Value of the term (y)
1	3
2	
3	
4	
5	19

Equation: $y =$ ______ x ___ ______

 ISBN: 9780170451468

Discrete applications

1 Kauri is selling bags of walnuts for $4 each at the market. His mum gives him six $1 coins so he will have change if people need it. Complete the table and graph to show how much money Kauri will have after selling x bags of walnuts.

Bags (x)	Money (y)
1	
2	
3	
4	
5	

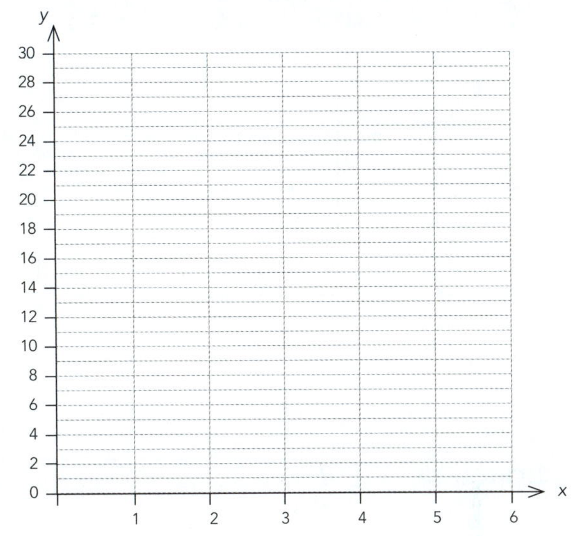

Equation: $y =$ ______ $x +$ ______

How much money will Kauri have when he has sold 12 bags of walnuts?

____

2 Ali owes her mother $135. She plans to pay her mum $15 every Friday until she has paid off the debt. Complete the table and graph to show how much Ali owes after x weeks.

Weeks (x)	Amount owed (y)
1	
2	
3	
4	
5	

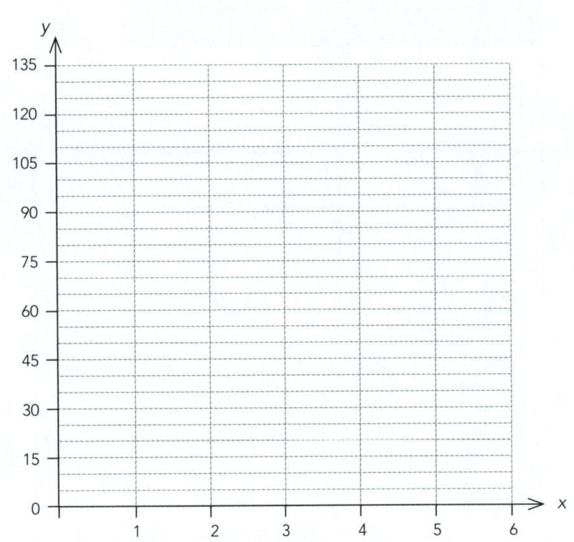

Equation: $y =$ ______ $x +$ ______

How much will Ali owe her mother after 8 weeks?

____

3 Scarlett is making batches of pancakes to feed her school camp. She needs 6 cups of flour for each batch. Complete the table and graph to show how many cups Scarlett will need for x number of batches.

Batch (x)	Cups of flour (y)
1	
2	
3	
4	
5	

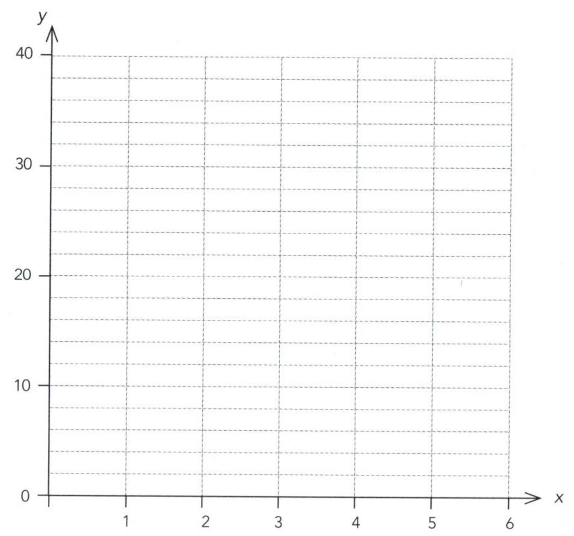

Equation: $y =$ ______________

How many cups of flour would Scarlett need for 12 batches of pancakes?

____

4 Every day, Josef and his sister have a vitamin tablet before they go to bed. There are 34 tablets left in the jar and each day they take a tablet each. Complete the table and graph to show the number of tablets left in the jar after x days.

Days (x)	Tablets (y)
1	
2	
3	
4	
5	

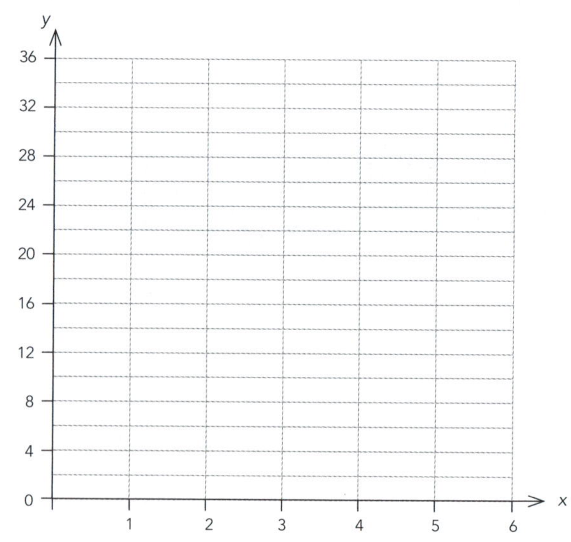

Equation: $y =$ ______________

How many tablets will be left in the jar after 13 days?

____

 ISBN: 9780170451468

Plotting continuous patterns

- **Continuous** data is obtained by **measuring**: it includes decimals or fractions.
- **Continuous** data is plotted on graphs as **lines**.
- Use a table to find **at least three points** and connect these with a **ruled line**.

Examples:

1 Plot the line given by the equation $y = 3x - 2$.

Step 1: Fill in the table.

x	Calculation	y	Coordinates
0	3 x 0 – 2	–2	(0, –2)
1		1	(1, 1)
2		4	(2, 4)
3		7	(3, 7)
4		10	(4, 10)
5		13	(5, 13)
6	3 x 6 – 2	16	(6, 16)

+3 +3 +3 +3 +3 +3

Sometimes you will get points that don't fit on the axes you are given. If this happens, just leave the point out.

Check term 6.
$y = 3 \times 6 - 2 = 16$ ✓

Step 2: Plot the points.

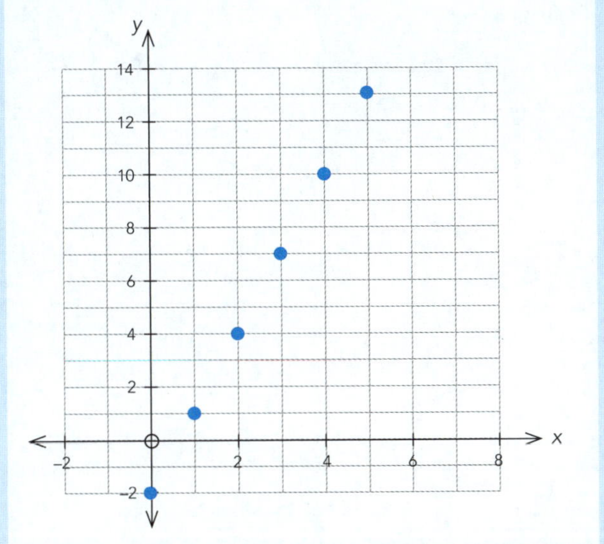

Step 3: Join the points with a **ruled** line.

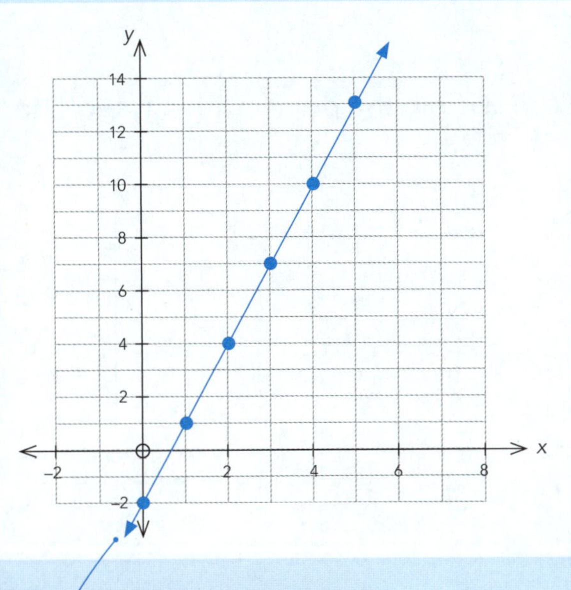

Notice the line continues beyond the coordinates and the axes.

2 Plot the line given by the equation $y = -2x + 8$.

Step 1: Fill in the table.

x	Calculation	y	Coordinates
0	−2 x 0 + 8	8	(0, 8)
1		6	(1, 6)
2		4	(2, 4)
3		2	(3, 2)
4		0	(4, 0)
5		−2	(5, −2)
6	−2 x 6 + 8	−4	(6, −4)

−2 −2 −2 −2 −2 −2

Check term 6.
$y = -2 \times 6 + 8 = -4$ ✓

Step 2: Plot the points.

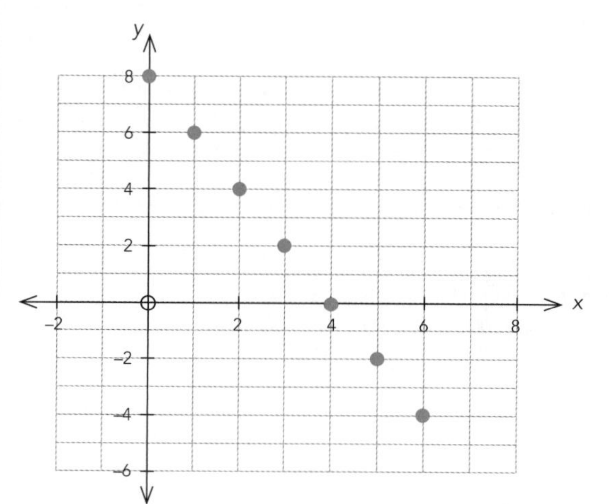

Step 3: Join the points with a **ruled** line.

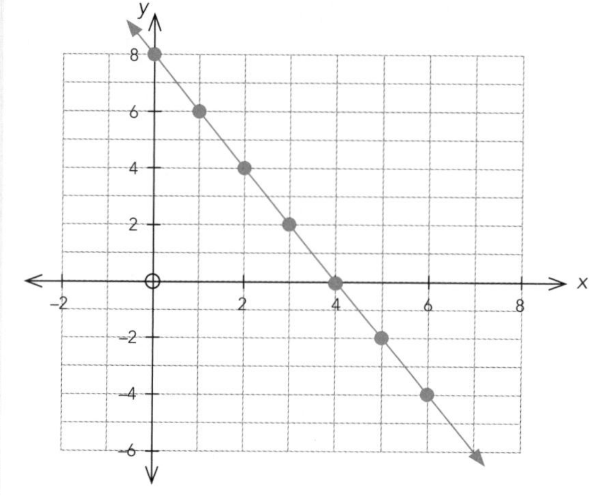

 ISBN: 9780170451468

Use the equation to complete the table, plot the points on the graph and draw a line representing the equation.

1 $y = 2x + 4$

x	Calculation	y	Coordinates
0			
1		6	(1, 6)
2			
3			
4			
5			
6			

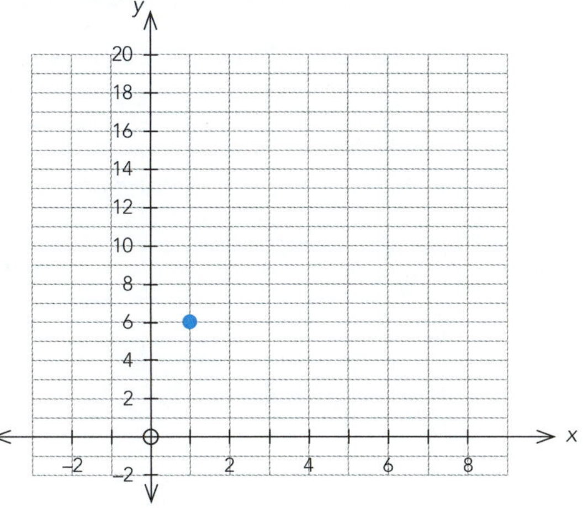

2 $y = -x + 15$

x	Calculation	y	Coordinates
0			
1			
2			
3			
4			
5			
6			

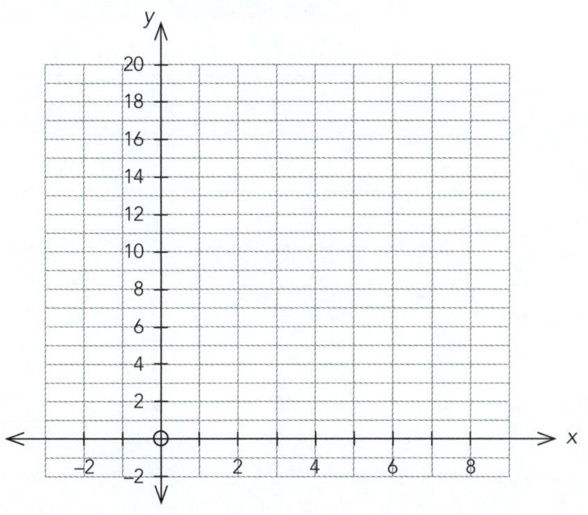

3 $y = 3x - 5$

x	y
0	
1	
2	
3	
4	
5	
6	

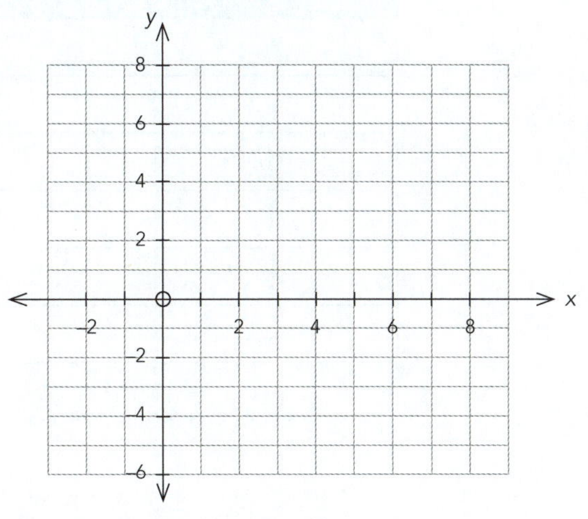

ISBN: 9780170451468

4 $y = -2x + 7$

x	y
0	
1	
2	
3	
4	
5	
6	

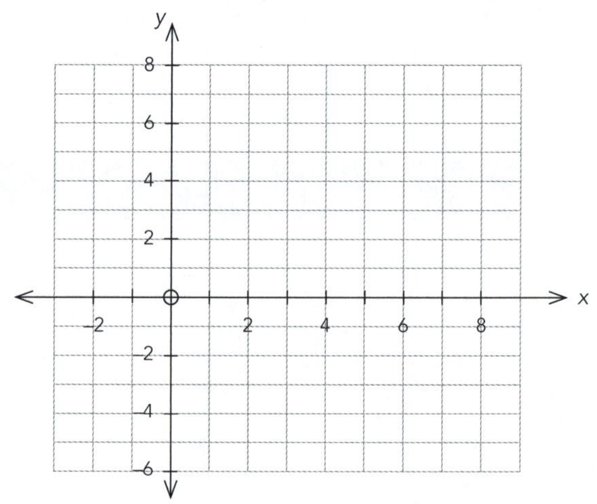

5 $y = 4x$

x	y
0	
1	
2	
3	
4	
5	
6	

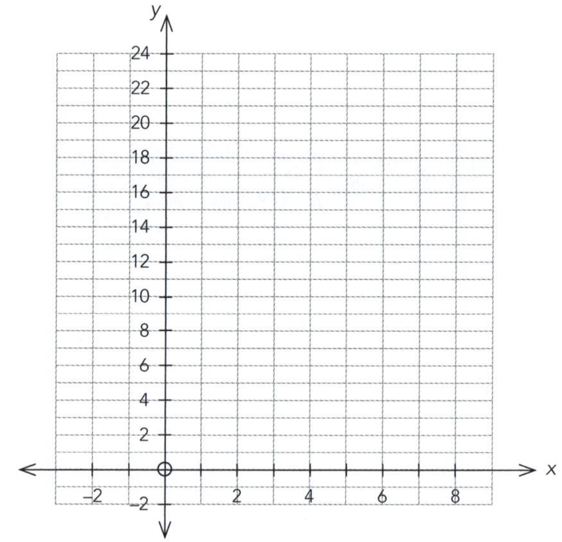

6 $y = -2x$

x	y
0	
1	
2	
3	
4	
5	
6	

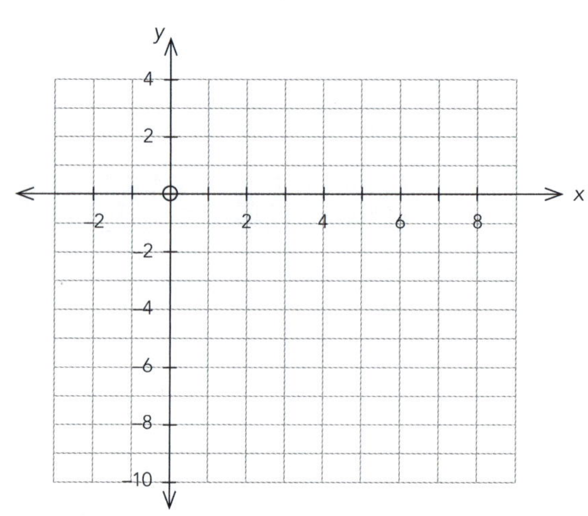

ISBN: 9780170451468

7 $y = -x - 1$

x	y
0	
1	
2	
3	
4	
5	
6	

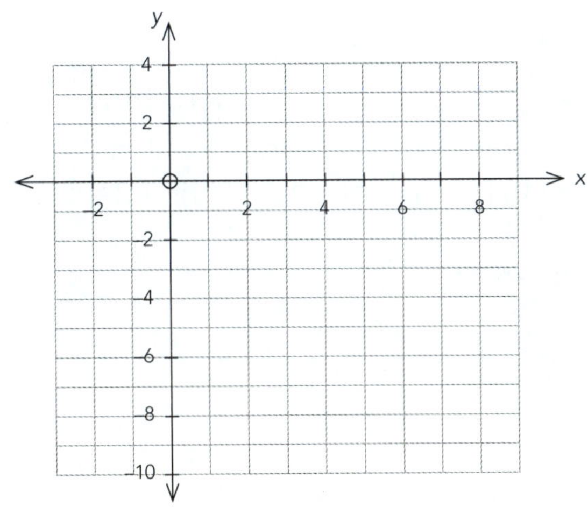

8 $y = -3x + 1$

x	y
0	
1	
2	
3	
4	
5	
6	

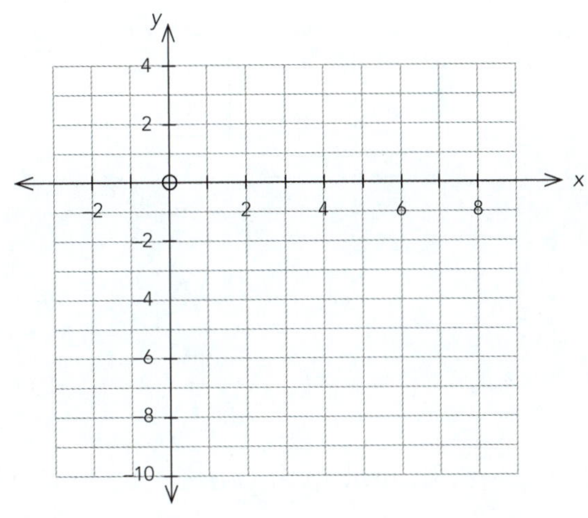

9 $y = -2x - 3$

x	y
0	
1	
2	
3	
4	
5	
6	

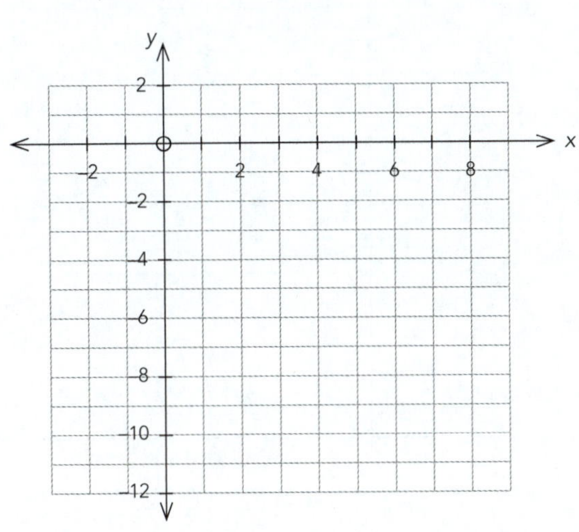

ISBN: 9780170451468

Finding linear equations from graphs

- Look where the line passes **exactly through lattice points** and use **at least three consecutive points** to fill in the table and write the equation.
- Lattice points are on the intersections of gridlines.

Example:

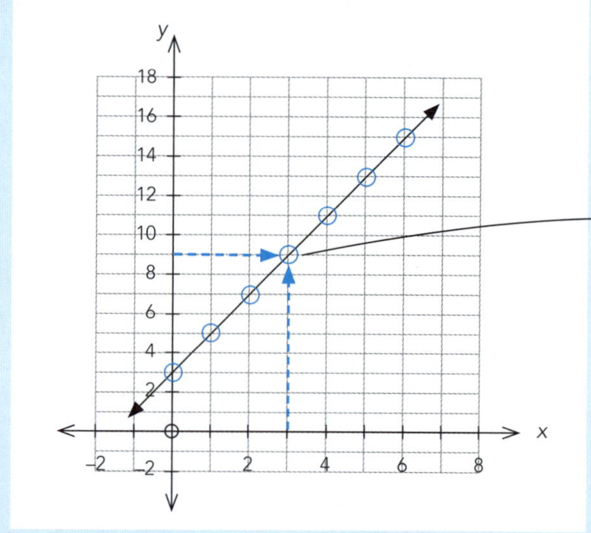

x	y
0	**3**
1	5
2	7
3	**9**
4	11
5	13
6	15

y = <u>2</u> x term number + <u>3</u>

Equation: $\mathbf{y = 2x + 3}$

Complete the table and use it to find the equation of the line.

1

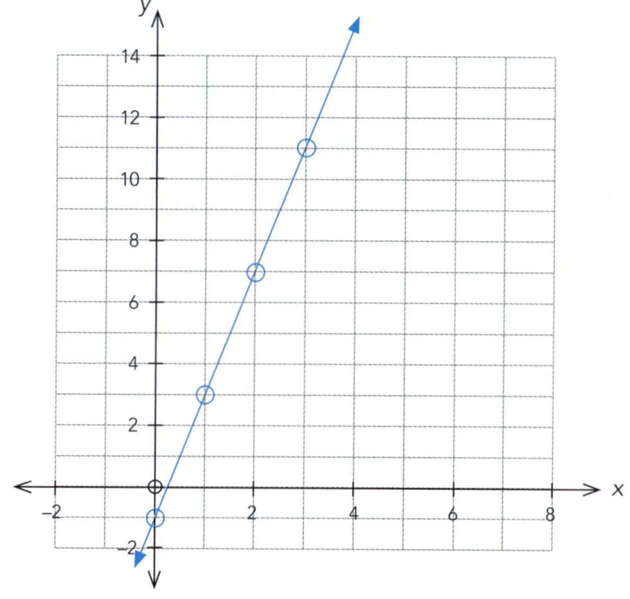

x	y
0	
1	
2	
3	
4	

Equation: $y =$ ______________

 ISBN: 9780170451468

2

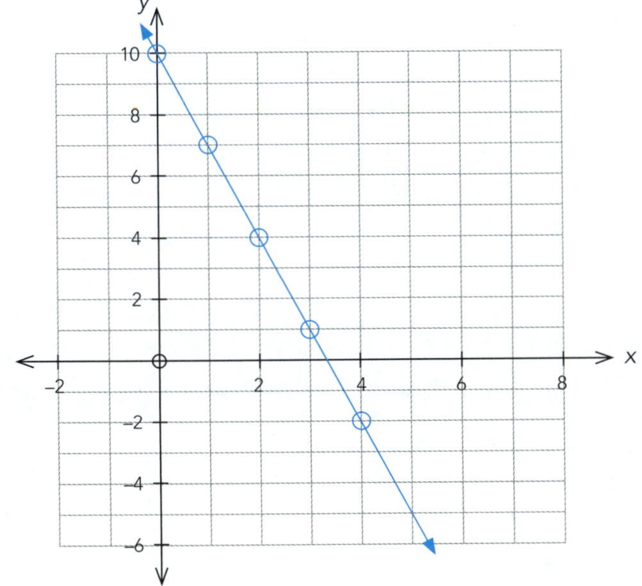

x	y
0	
1	
2	
3	
4	

Equation: $y =$ __________

3

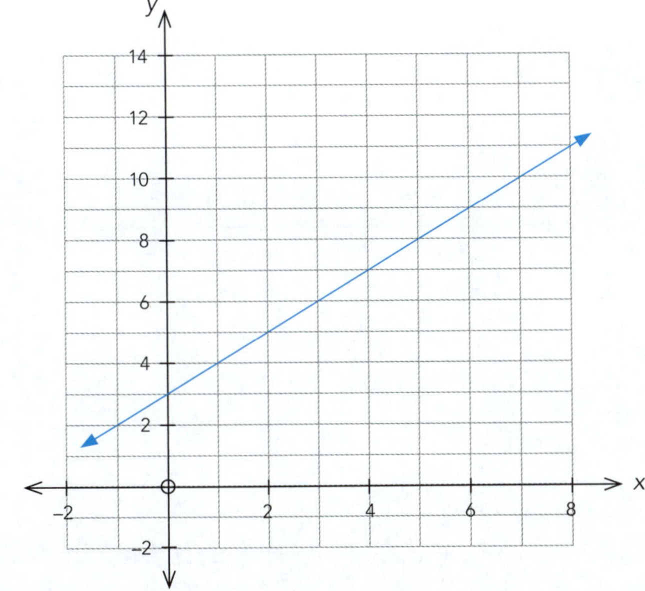

x	y
0	
1	
2	
3	
4	

Equation: $y =$ __________

4

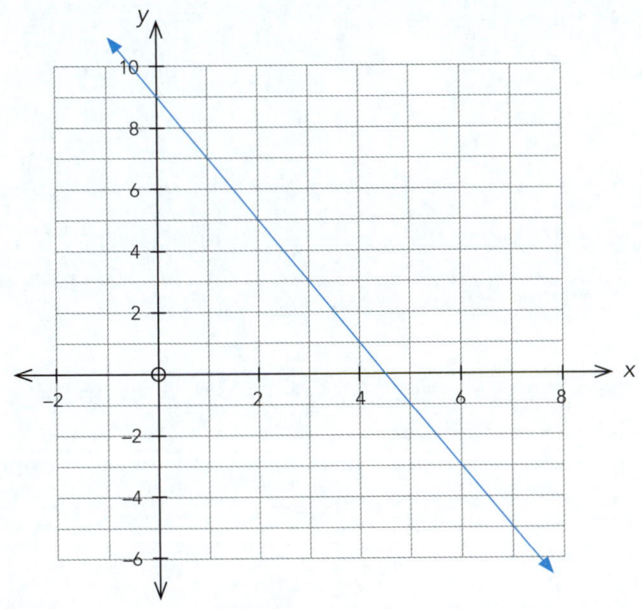

x	y
0	
1	
2	
3	
4	

Equation: $y =$ __________

ISBN: 9780170451468

Horizontal and vertical lines

- Horizontal and vertical lines have only **one letter** in their equation.
- Be very careful drawing these — it's easy to get them the wrong way round.
- You can draw up a table, or just write a list of **at least three** points.

Examples:

1 $y = 4$

x	y	Coordinates
0	4	(0, 4)
1	4	(1, 4)
2	4	(2, 4)
3	4	(3, 4)

It doesn't matter what the x value is, y will **always be 4**.

Plot at least three points, then join the points.

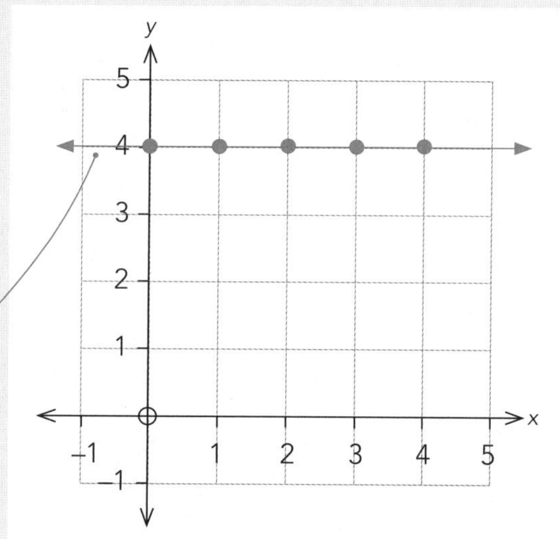

Equations with **y = …** are always **horizontal** lines.

2 $x = -2$

It doesn't matter what the y value is, x will **always be –2**.

Step 1: List at least three points where $x = -2$, e.g. (–2, 0), (–2, 1), (–2,2), (–2, 3).
Step 2: Plot the points, then join them.

Equations with **x = …** are always **vertical** lines.

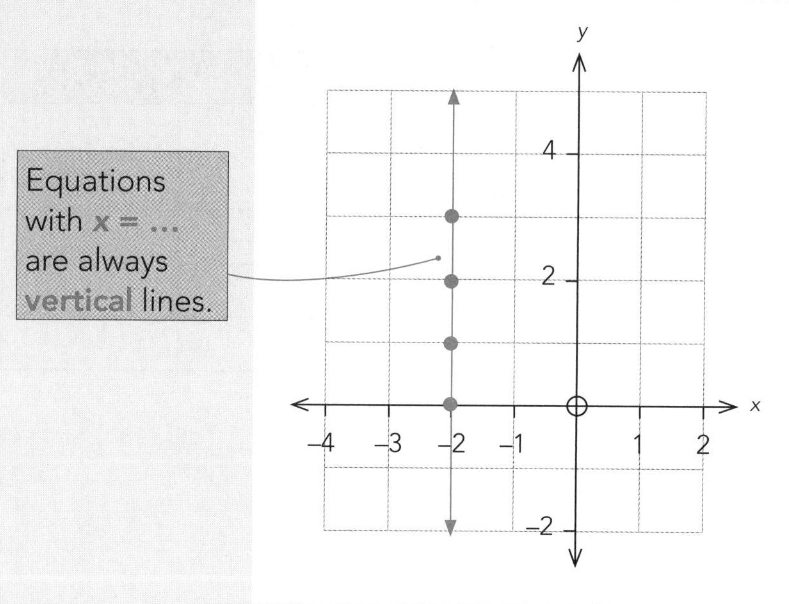

 ISBN: 9780170451468

Highlight/circle the correct equation for each graph.

1

$x = 1$

$y = -1$

$x = -1$

$y = 1$

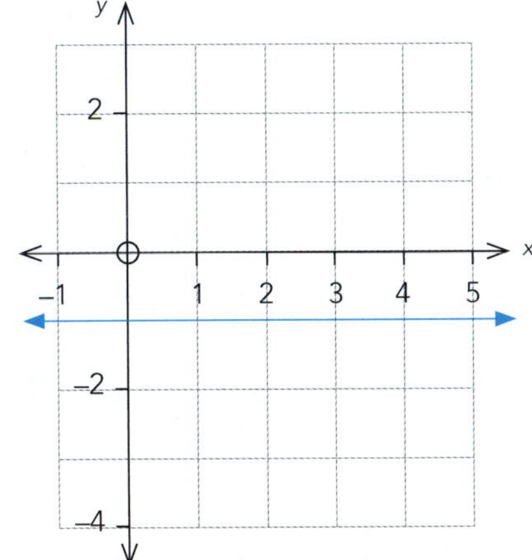

2

$x = -2$

$y = -2$

$x = 2$

$y = 2$

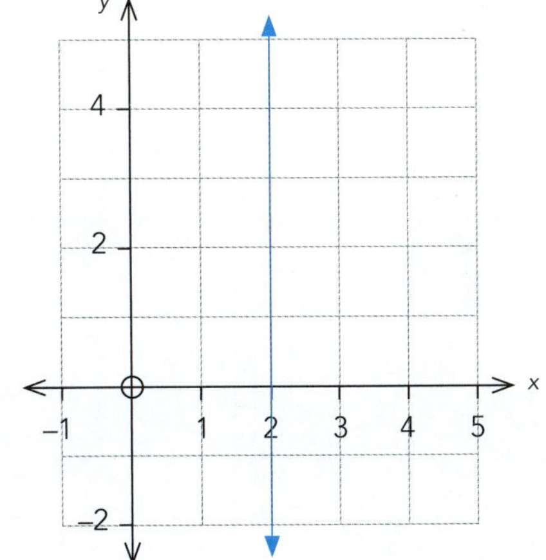

3

$x = -0.5$

$y = \frac{1}{2}$

$x = \frac{1}{2}$

$y = -0.5$

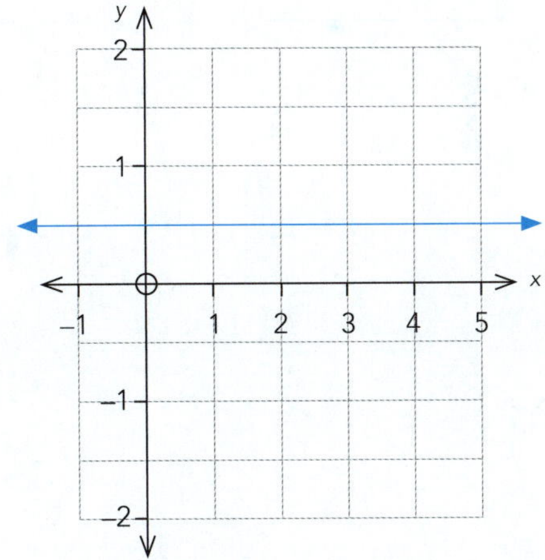

4

$x = \frac{3}{2}$

$y = -1.5$

$x = -1.5$

$y = -\frac{3}{2}$

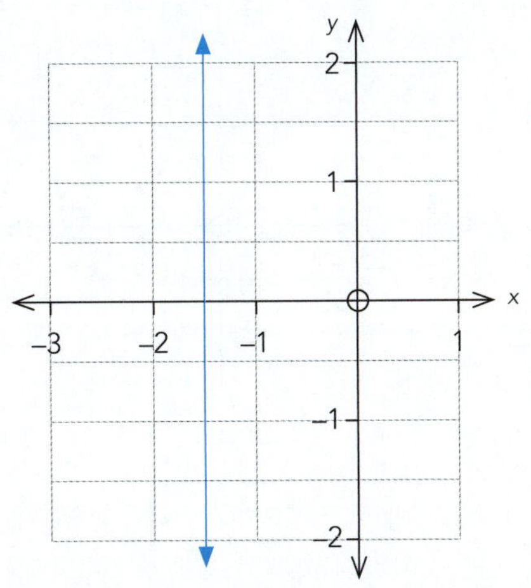

ISBN: 9780170451468

Draw these lines on the graphs.

5 $y = 3$

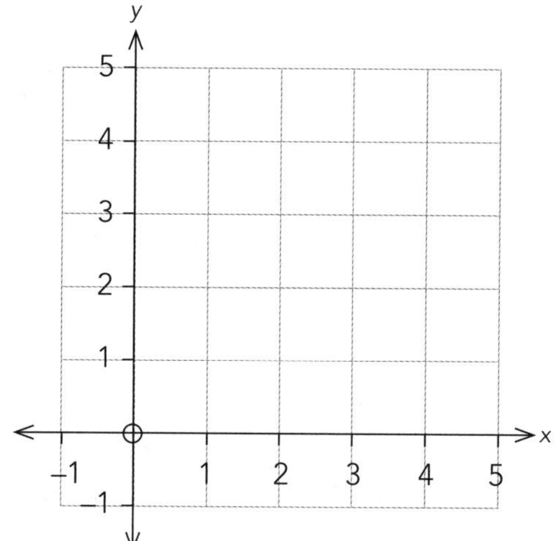

6 $x = 2$

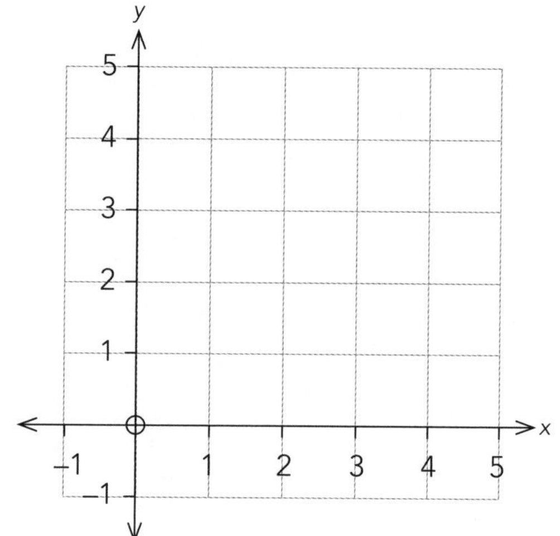

7 $x = -1$

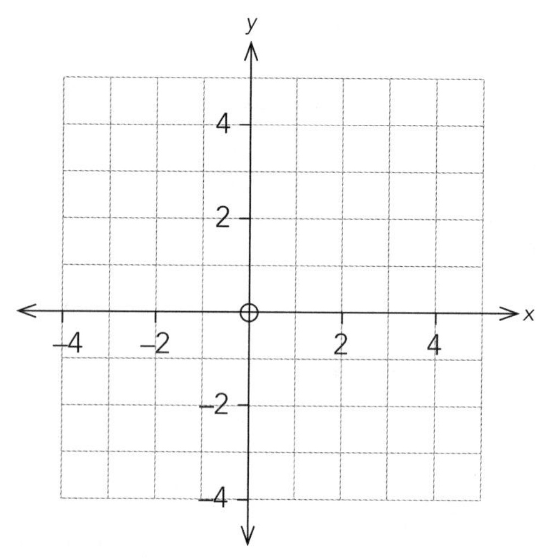

8 $y = -3$

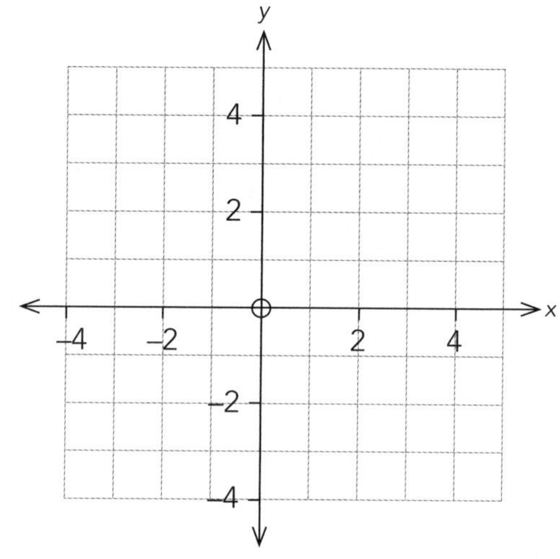

9 $x = 0$

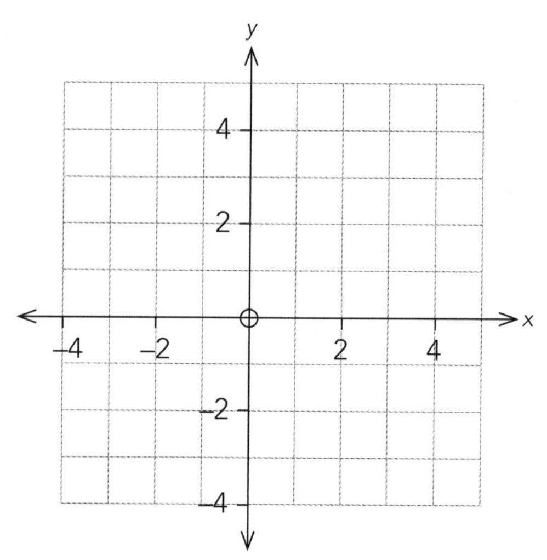

10 $y = 0$

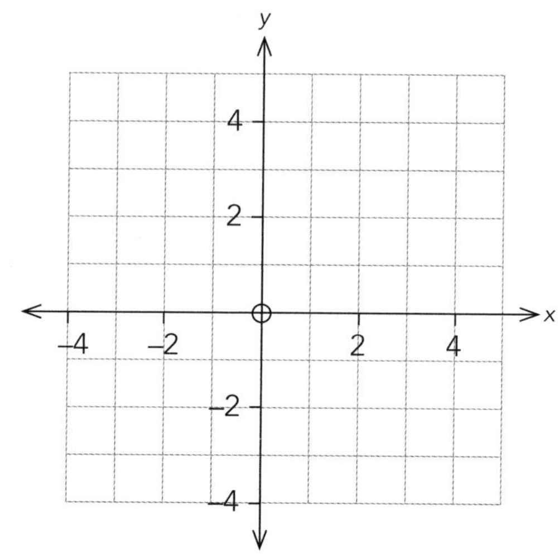

ISBN: 9780170451468

The gradient of a line

- The gradient is the **steepness**, or **slope**, of a line.
- The gradient is usually represented by the letter **m**.

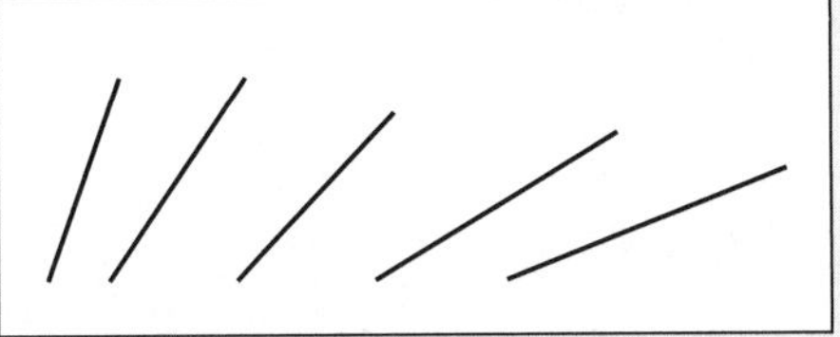

These lines all have **positive** gradients.

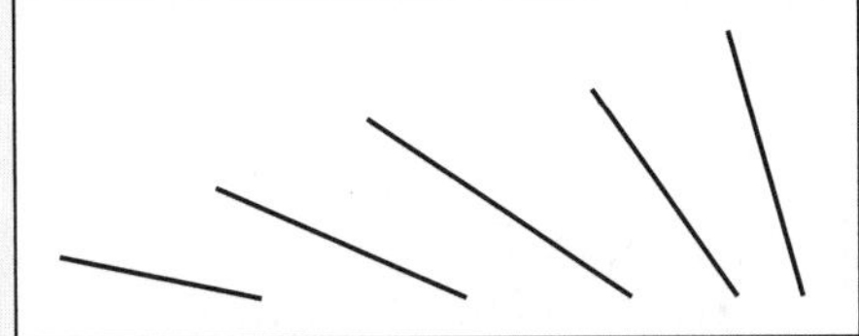

These lines all have **negative** gradients.

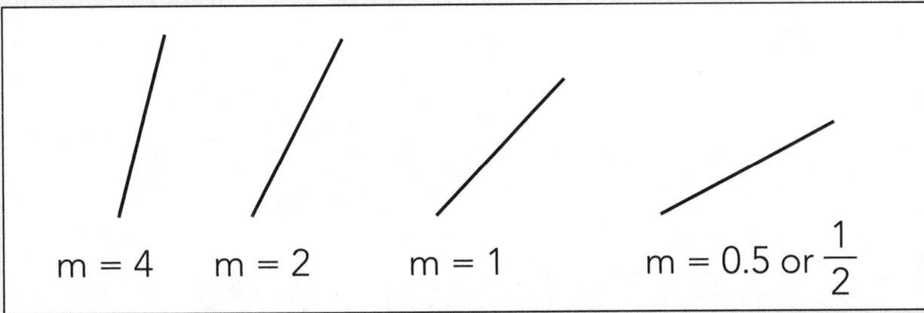

The steeper the line, the bigger the gradient.

The gradient is calculated using the formula $m = \frac{\text{change in } y}{\text{change in } x}$ or $\frac{\text{rise}}{\text{run}}$.

Examples:

1

Step 1: Pick two points where the line goes **through lattice points on the grid**.

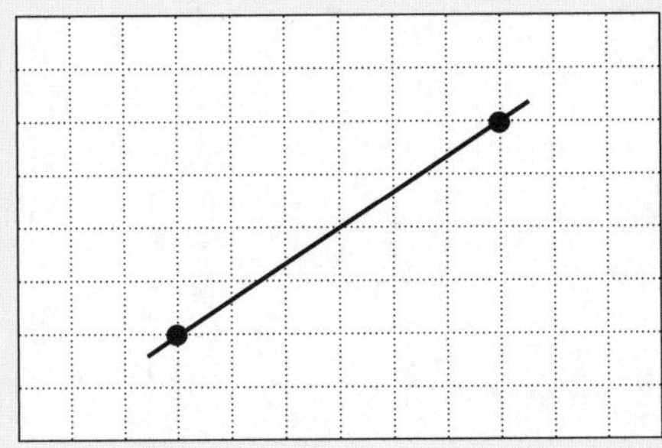

Step 2: Draw a right-angled triangle connecting these points and count the number of squares up (**rise**) and across (**run**).

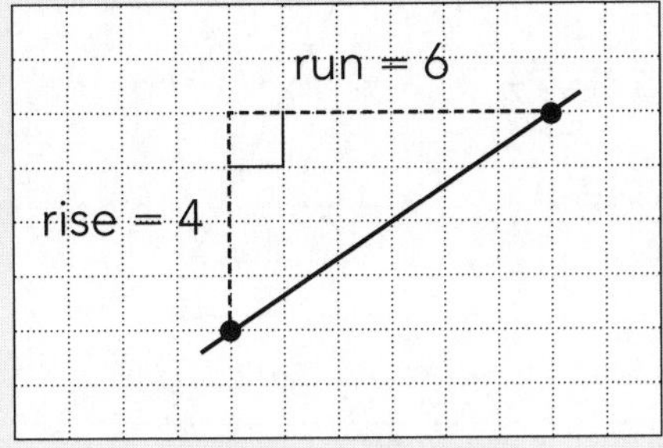

Step 3: Simplify if possible. $m = \frac{4}{6}$ or $\frac{2}{3}$ or $0.\dot{6}$

ISBN: 9780170451468

2

Step 1: Pick two points where the line goes through the intersections of the grid.

Step 2: Draw a right-angled triangle connecting these points and count the number of squares up (**rise**) and across (**run**).

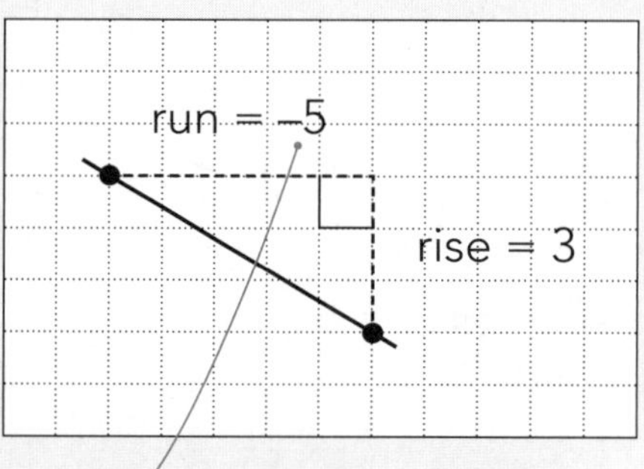

Step 3: $m = -\frac{3}{5}$

Notice we have moved to the **left**, so this is a **negative** 5.

3

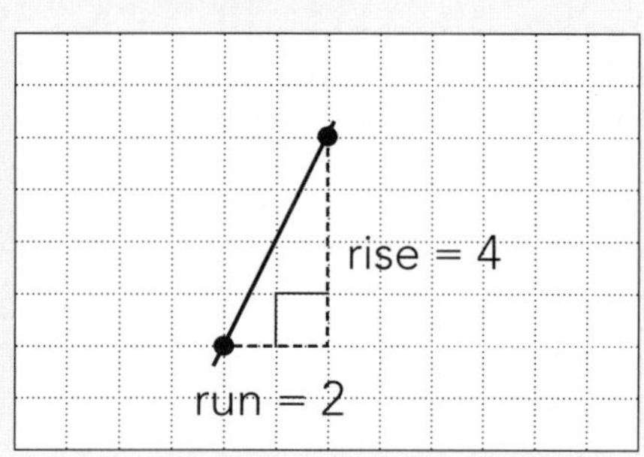

$$m = \frac{4}{2} \text{ or } \frac{2}{1} \text{ or } 2$$

Any of these answers is acceptable. Some gradients simplify to single digits.

4

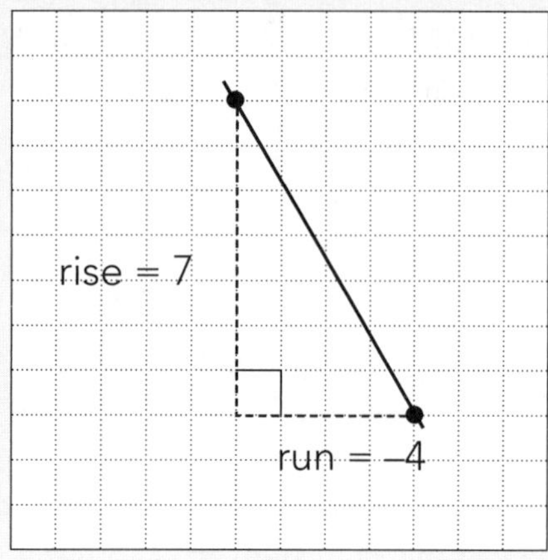

$$m = \frac{7}{-4} \text{ or } \frac{-7}{4} \text{ or } -1\frac{3}{4} \text{ or } -1.75$$

Improper fractions are the most convenient form for writing gradients.

Remember, the negative for a fraction can be written in any of these forms:

$$\frac{-5}{1} = -\frac{5}{1} = \frac{5}{-1}$$

However, fractions are not normally written with negative denominators.

 ISBN: 9780170451468

Find the gradients of these lines. Write your answers as fractions or whole numbers.

1

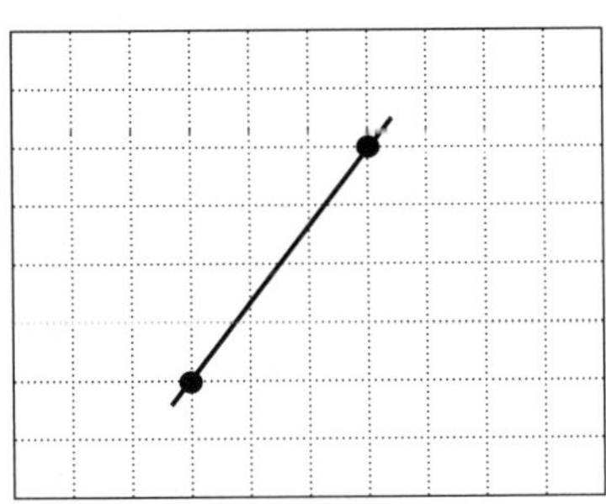

m = ________

2

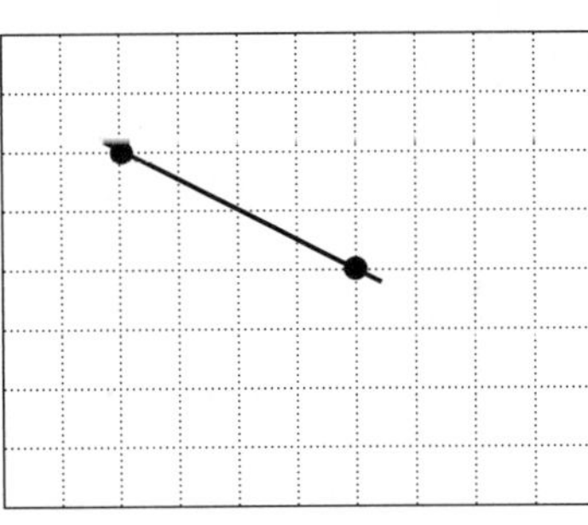

m = ________

3

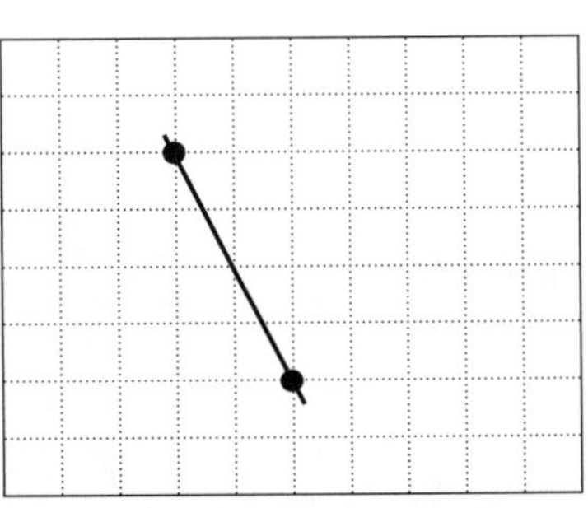

m = ________

4

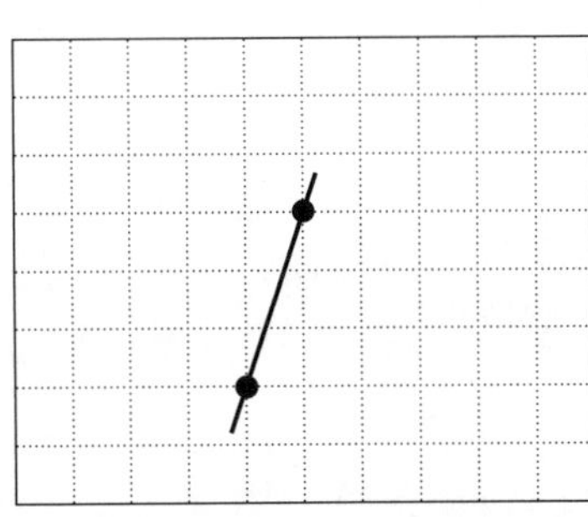

m = ________

5

m = ________

6

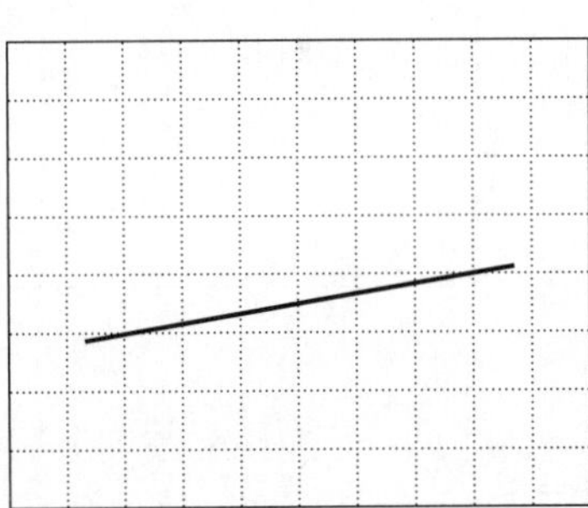

m = ________

7

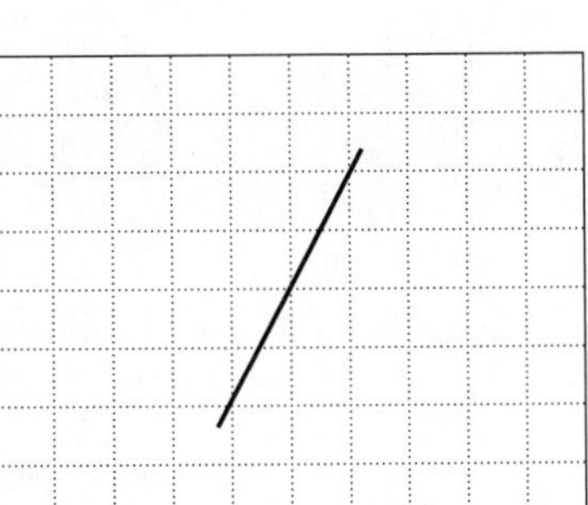

m = ________

8

m = ________

ISBN: 9780170451468

Draw line segments to show these gradients.

9 $m = \frac{5}{6}$

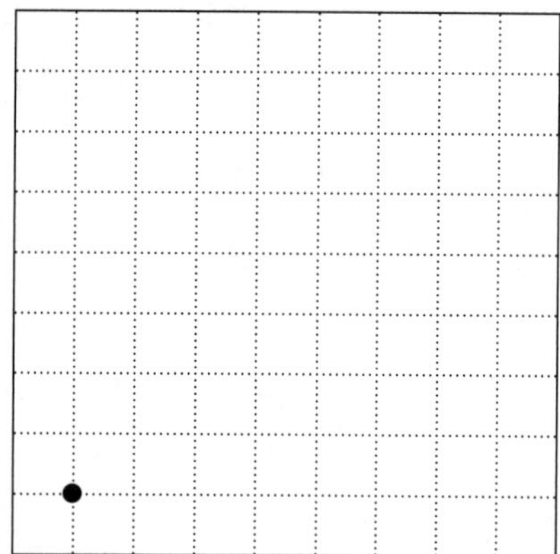

10 $m = -\frac{4}{5}$

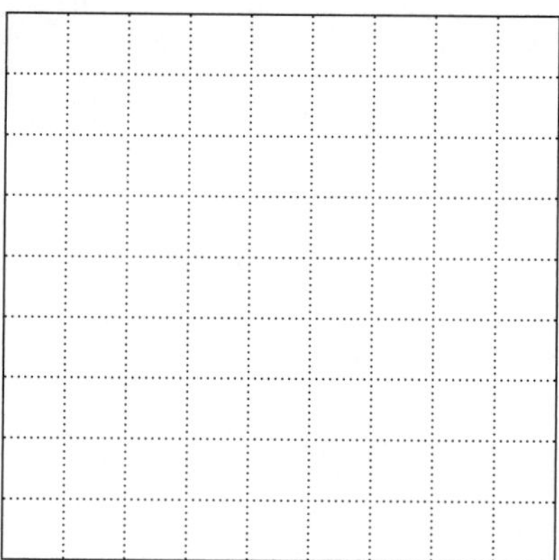

11 $m = \frac{7}{6}$

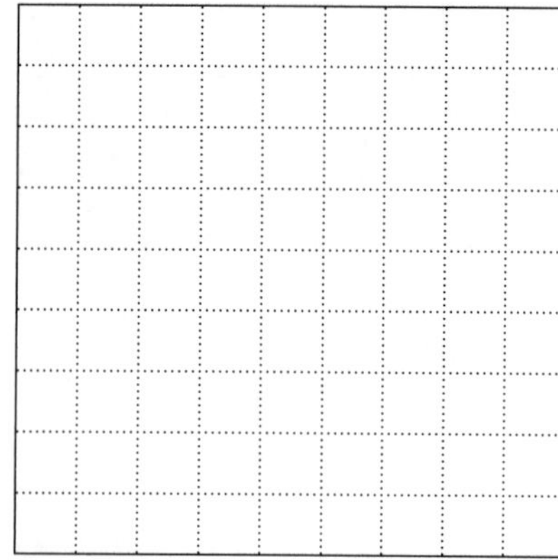

12 $m = -3$

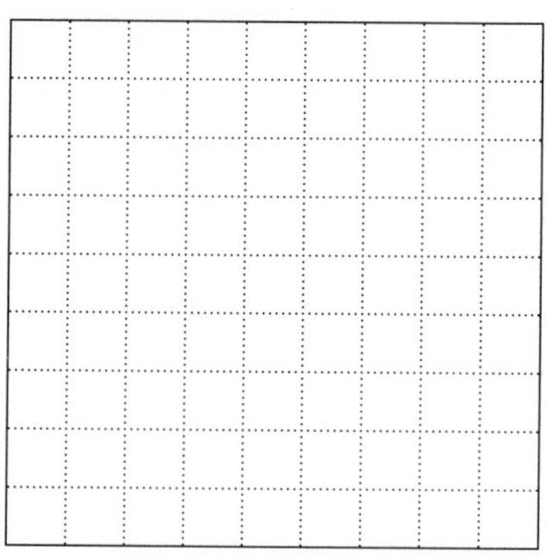

13 $m = 4$

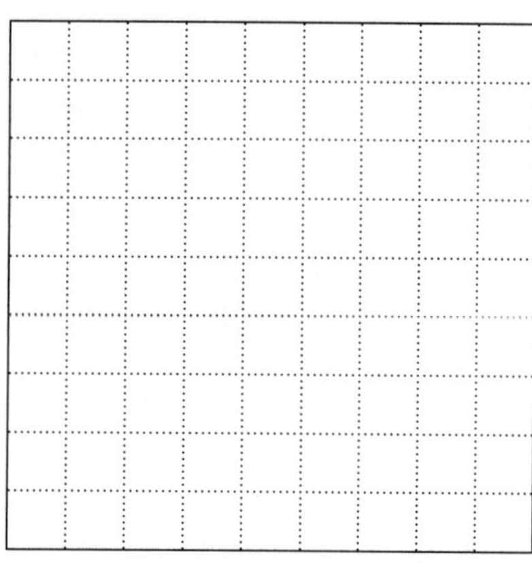

14 $m = 0.5$

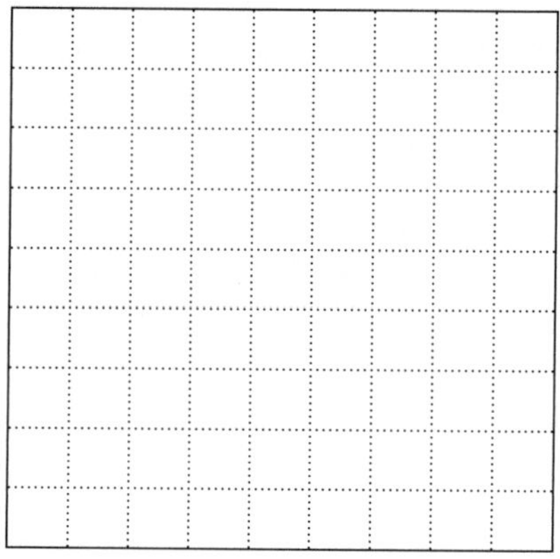

 ISBN: 9780170451468

Find the gradients of these lines. Write your answers as fractions or whole numbers.

15

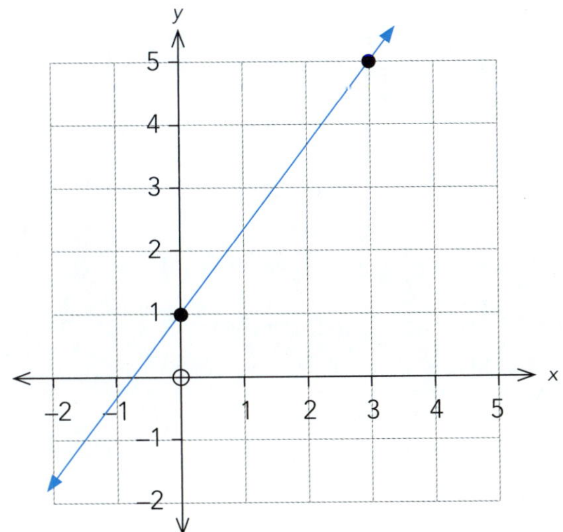

m = ________

16

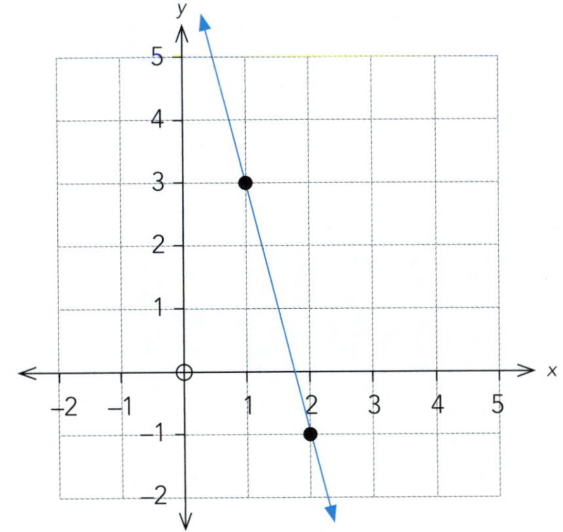

m = ________

17

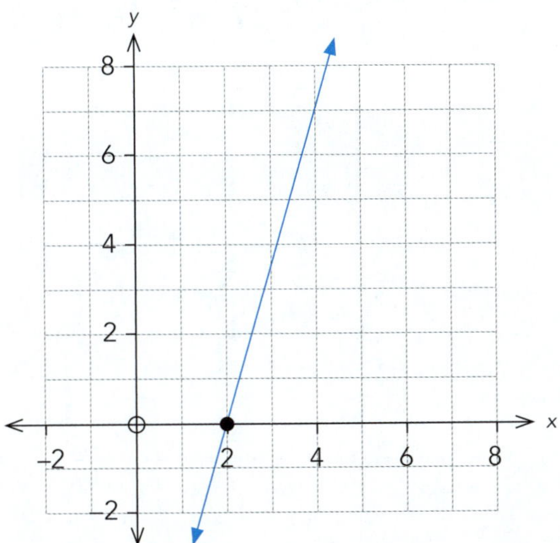

m = ________

18

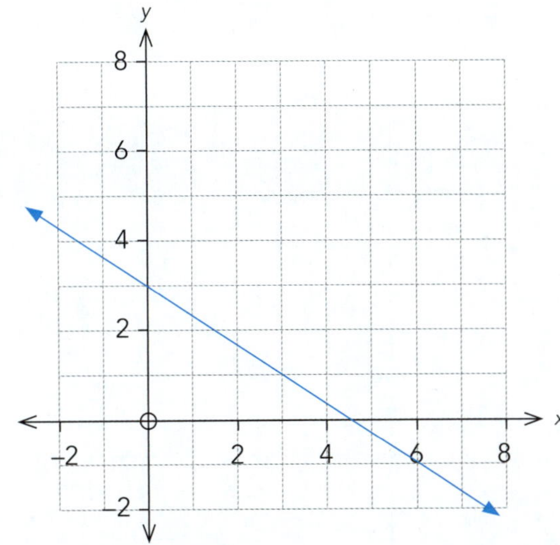

m = ________

19

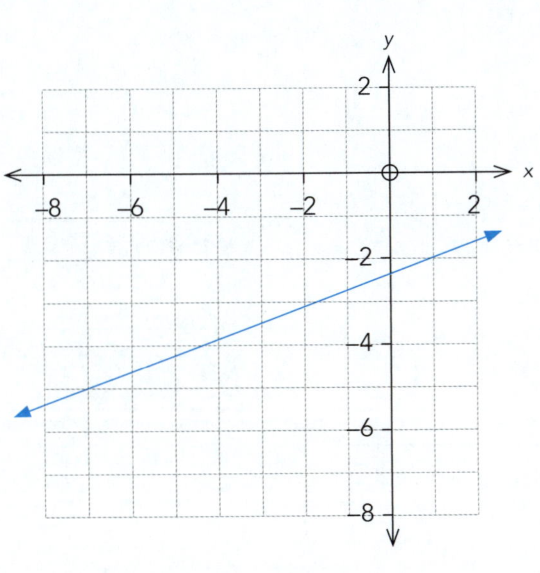

m = ________

20

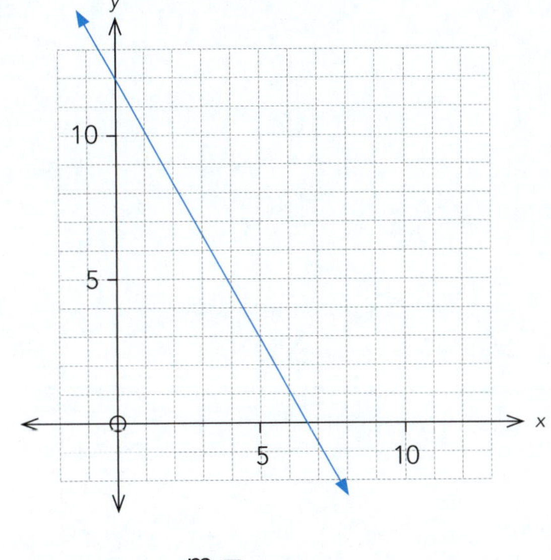

m = ________

Drawing straight lines using the gradient and *y* intercept

- When you are given an equation of a line, use the format $y = mx + c$.

$$y = mx + c$$

m is the coefficient of x.

$$m = \text{gradient} = \frac{\text{rise}}{\text{run}}$$

c = the *y* intercept

The point where the line cuts the *y*-axis and the value for *y* when we substitute $x = 0$ into the equation.

Examples:

1 Draw the graph of $y = \frac{2}{3}x + 1$.

Step 1: Plot the *y* intercept. $y = 1$

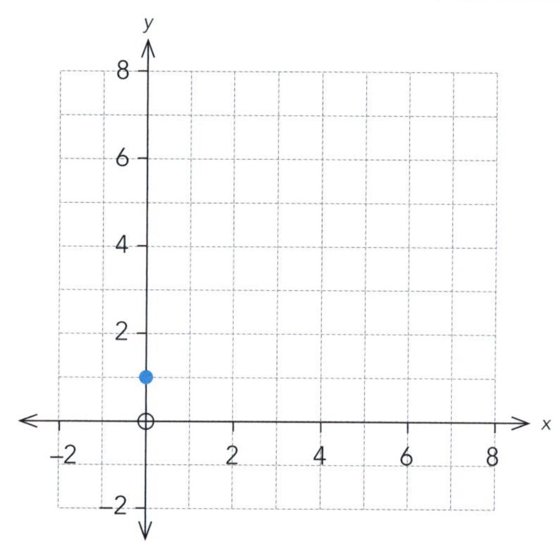

Step 2: From the **y intercept** (**not** the origin), use the gradient to plot **at least two more** points. $m = \frac{2}{3}$

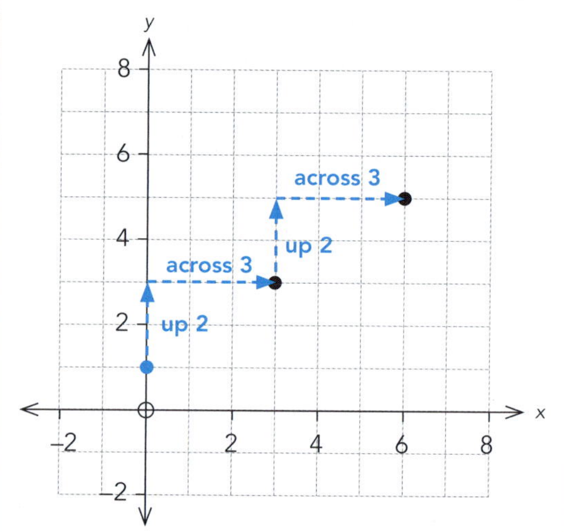

Step 3: Draw a **ruled** line through the points.

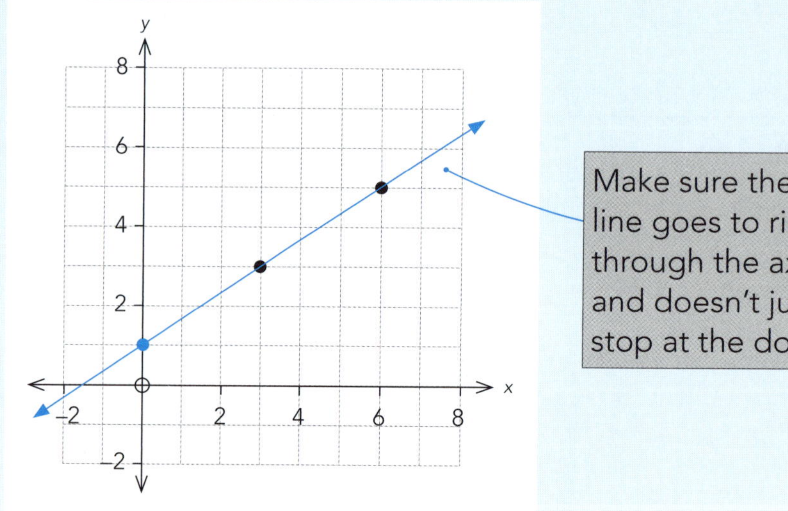

Make sure the line goes to right through the axis and doesn't just stop at the dots.

 ISBN: 9780170451468

2 Draw the graph of $y = \frac{-3}{4}x + 7$.

Step 1: Plot the *y* intercept. y = 7

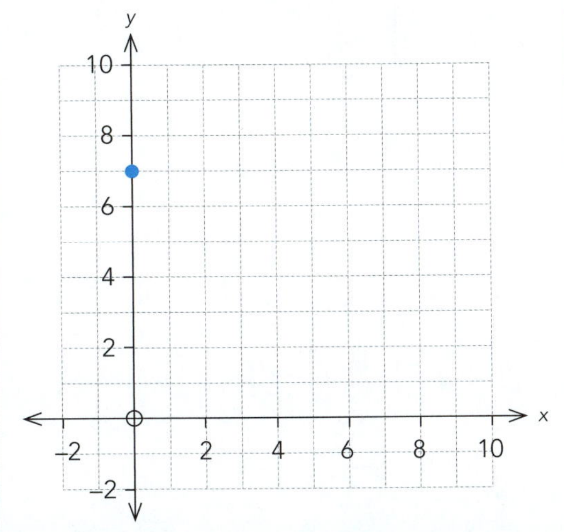

Step 2: From the **y intercept**, use the gradient to plot **at least two more** points. $m = \frac{-3}{4}$

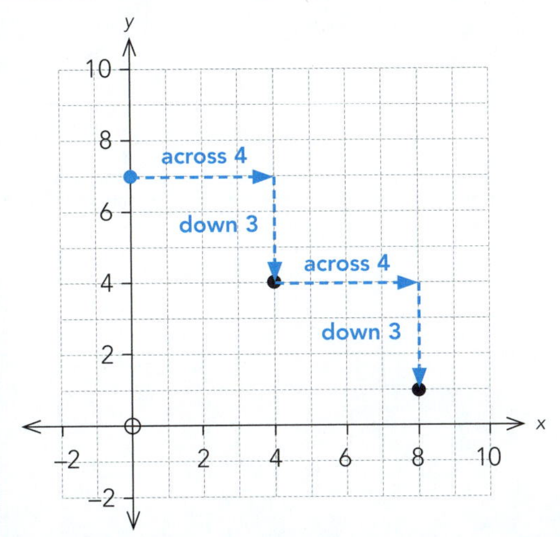

Step 3: Draw a **ruled** line through the points.

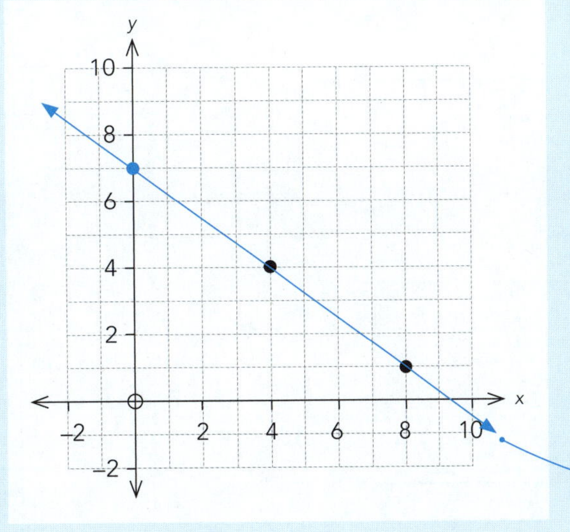

Make sure the line goes to right through the axis and doesn't just stop at the dots.

ISBN: 9780170451468

Match the equations with the graphs.

$y = \frac{4}{3}x - 1$	$y = -3x - 1$	$y = -\frac{5}{2}x + 4$	$y = \frac{9}{2}x - 2$	$y = -\frac{1}{3}x + 4$	$y = \frac{3}{4}x - 2$

1

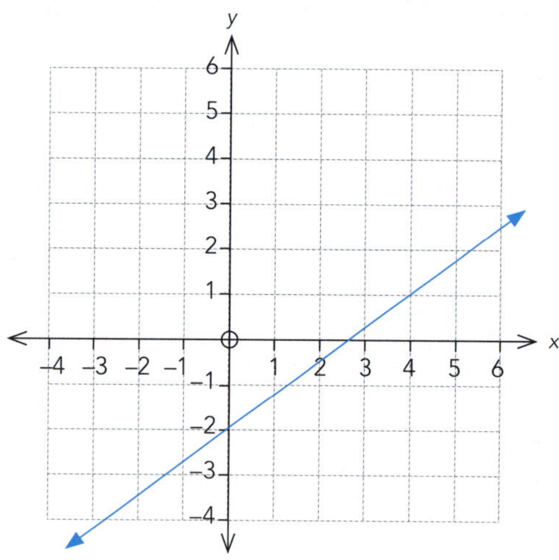

2

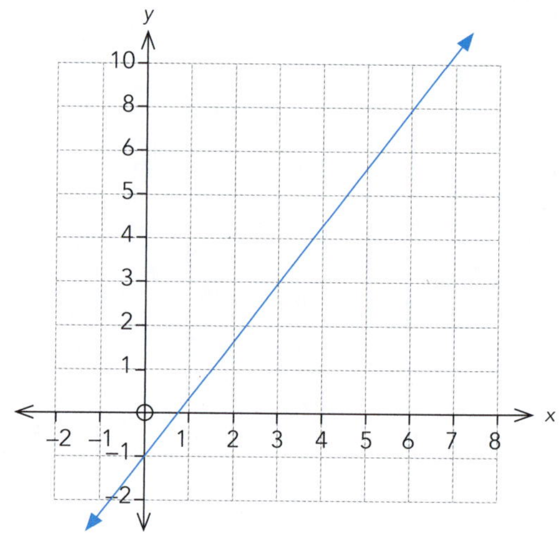

3

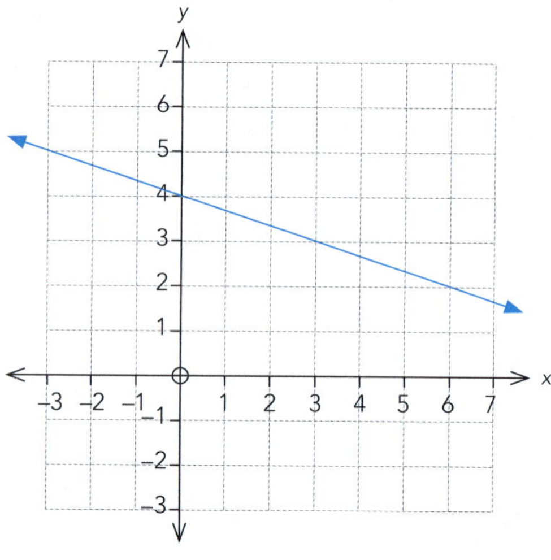

4

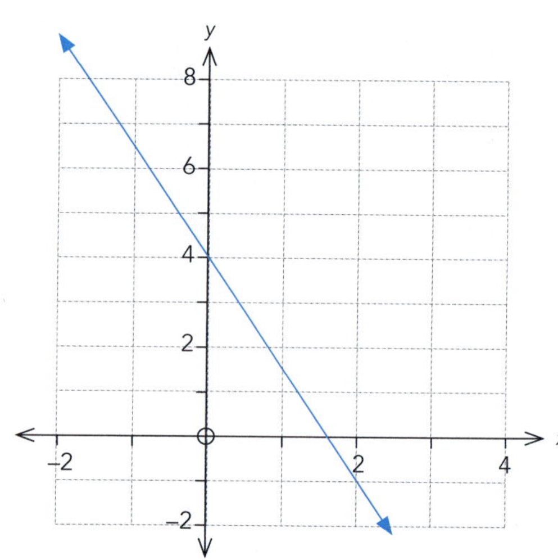

5

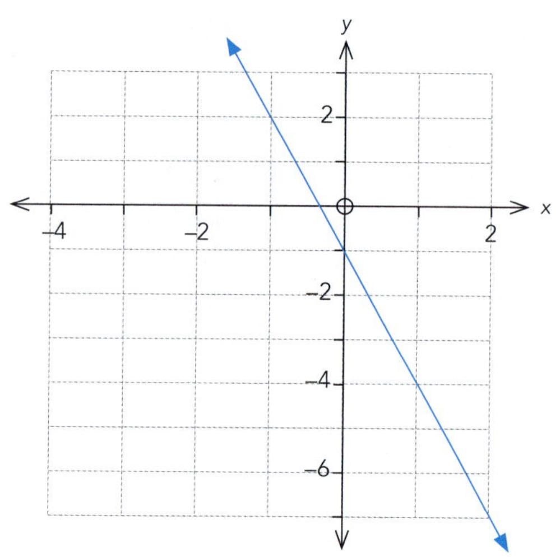

6

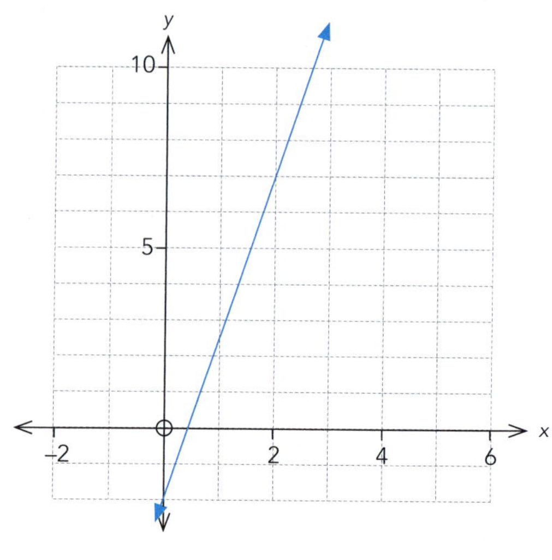

 ISBN: 9780170451468

Plot these lines on the axes using the y intercept and gradient.

7 $y = \frac{3}{2}x - 1$

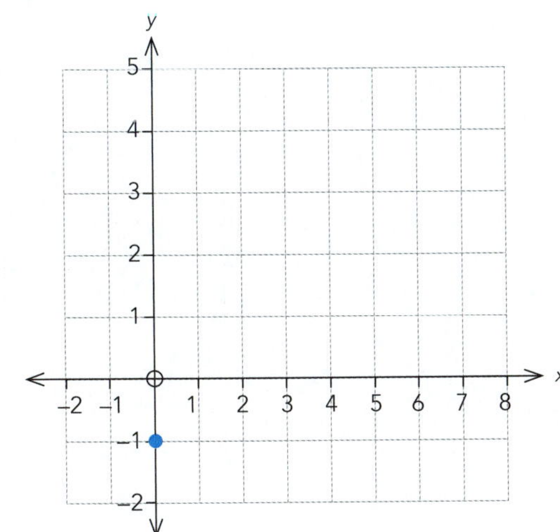

8 $y = -\frac{4}{1}x + 5$

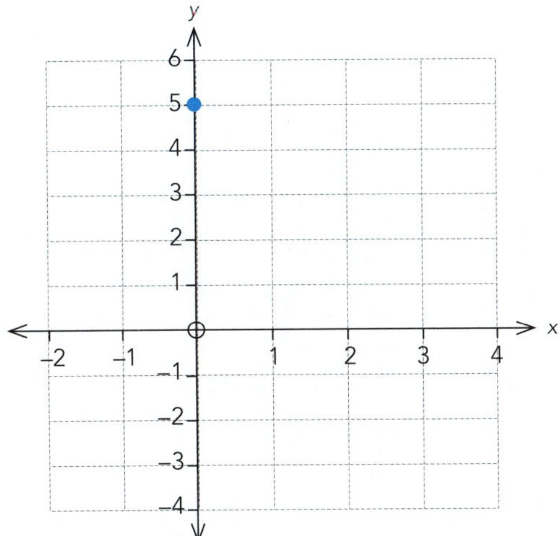

9 $y = \frac{1}{3}x - 4$

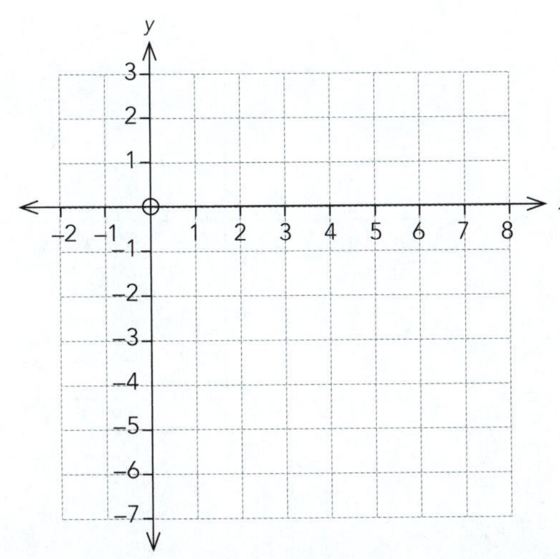

10 $y = \frac{5}{2}x - 3$

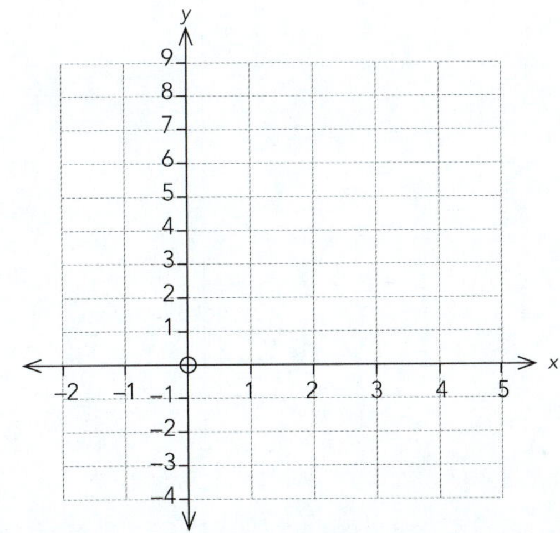

11 $y = \frac{-4}{3}x + 9$

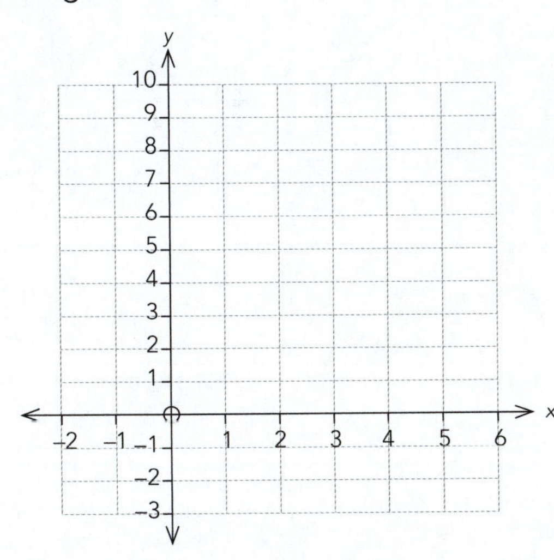

12 $y = \frac{2}{-3}x - 6$

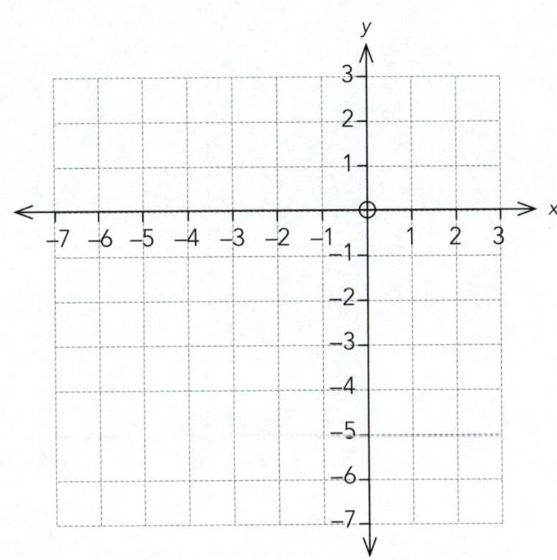

ISBN: 9780170451468

13 $y = \frac{1}{4}x + 1$

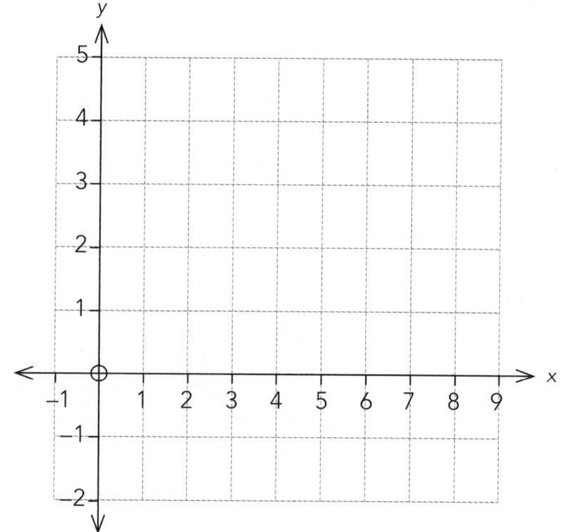

14 $y = -\frac{3}{2}x + 1$

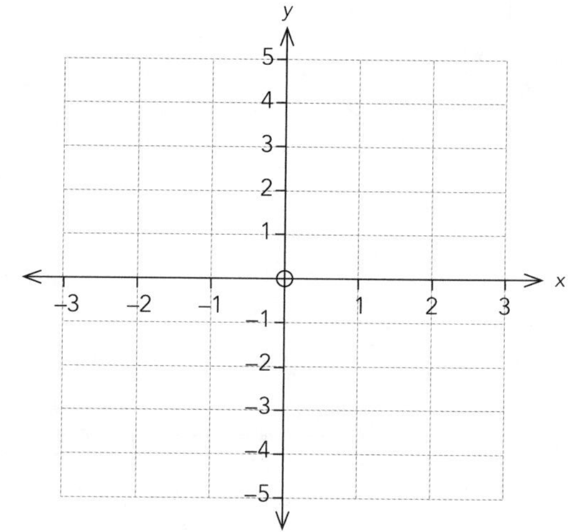

15 $y = \frac{5}{3}x$

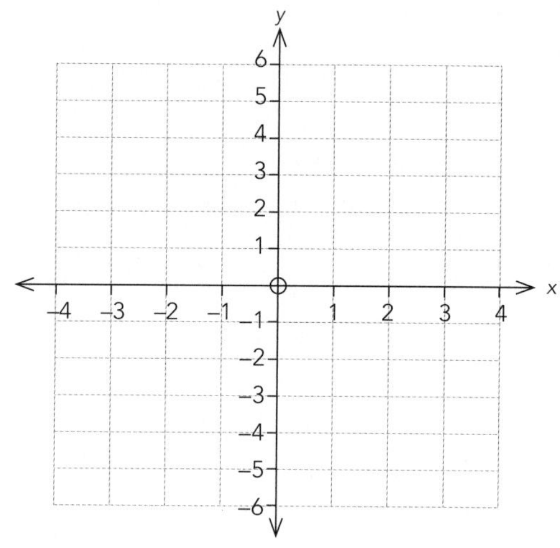

16 $y = -x$

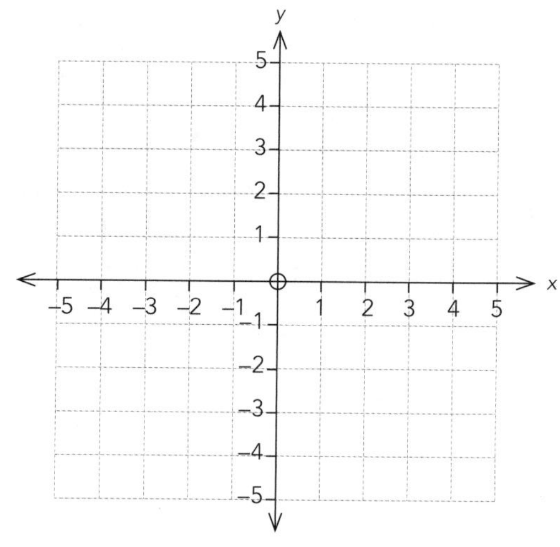

17 $y = 1.5x + 1$

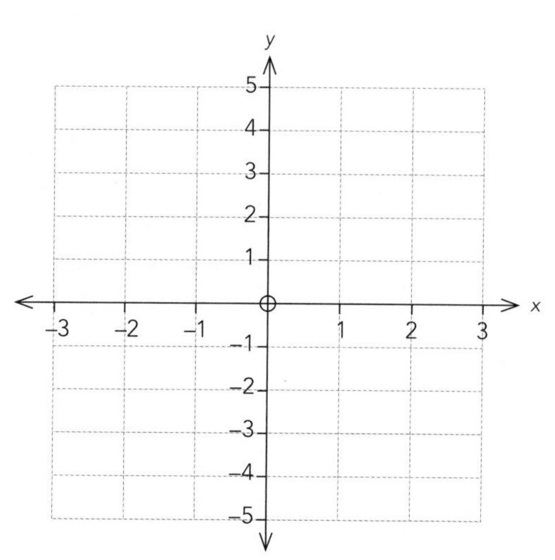

18 $y = -0.5x + 0.5$

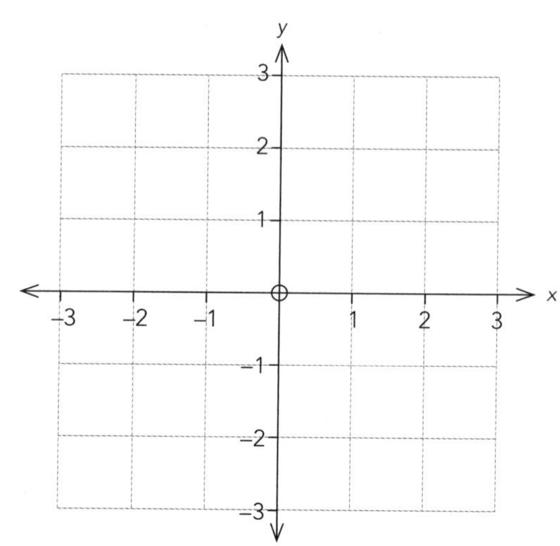

 ISBN: 9780170451468

Writing equations from a graph

- Use $y = mx + c$ to write the equations of lines.

Examples:

1 **Step 1:** Identify the y intercept.

Step 2: Use at least one point which passes through the intersections on the grid to calculate the gradient.

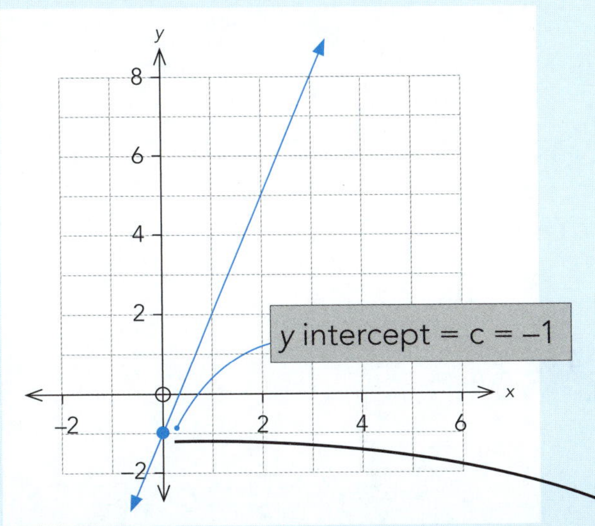

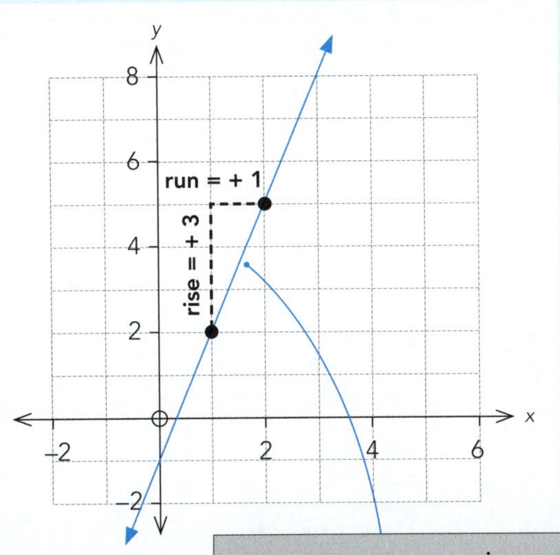

Step 3: Substitute into $y = mx + c$.

$$y = \frac{3}{1}x - 1 \quad \text{or} \quad y = 3x - 1$$

2 **Step 1:** Identify the y intercept.

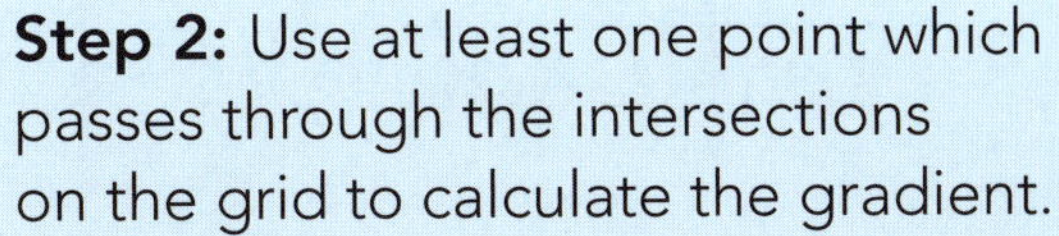

Step 2: Use at least one point which passes through the intersections on the grid to calculate the gradient.

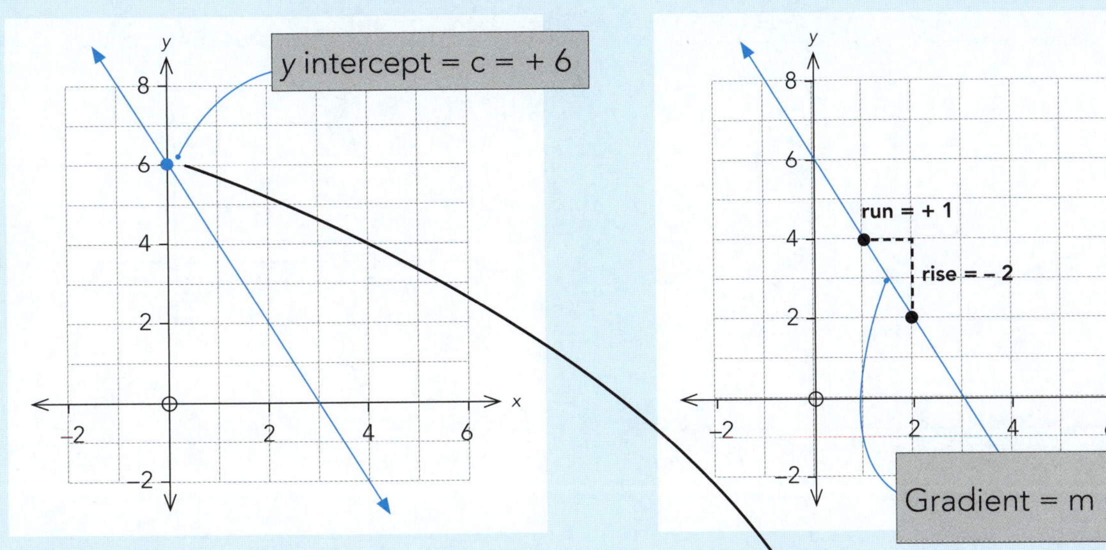

Step 3: Substitute into $y = mx + c$.

$$y = \frac{-2}{1}x + 6 \quad \text{or} \quad y = -2x + 6$$

ISBN: 9780170451468

Write the equations of these graphs lines.

1

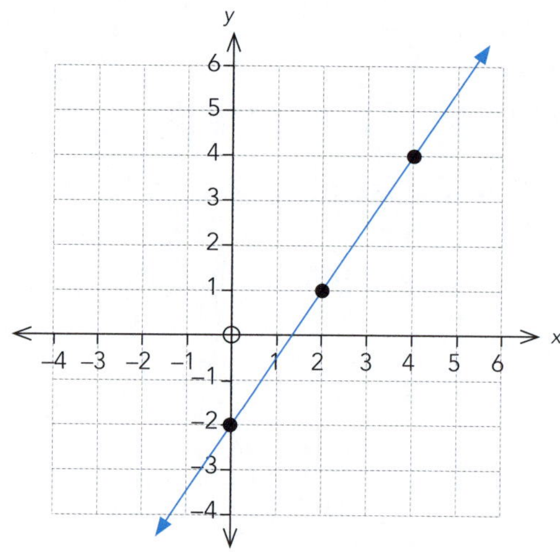

2

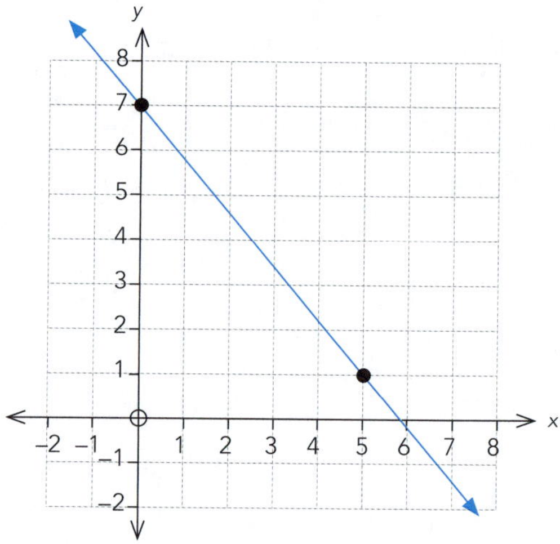

3

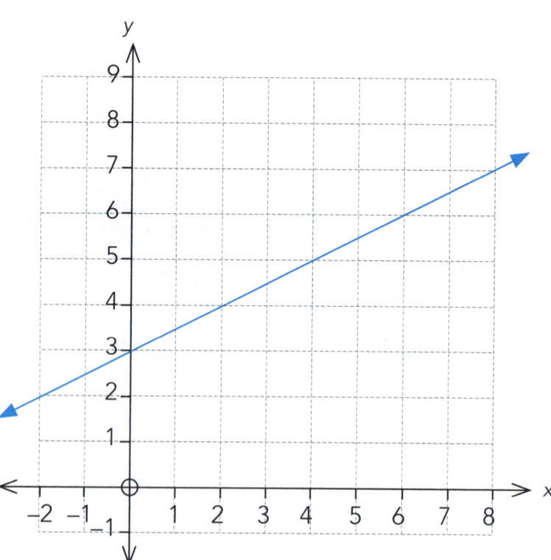

4

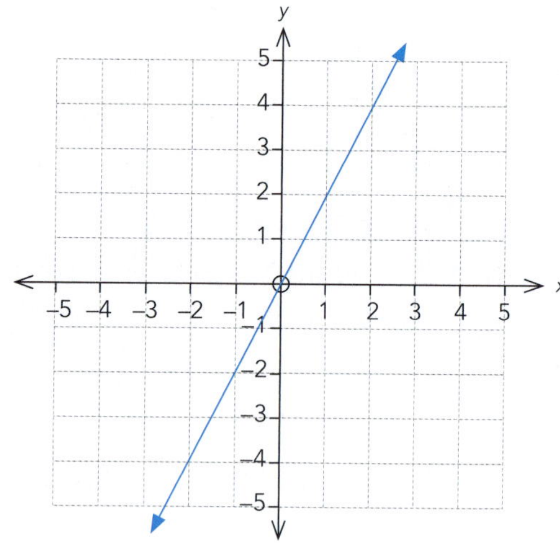

5

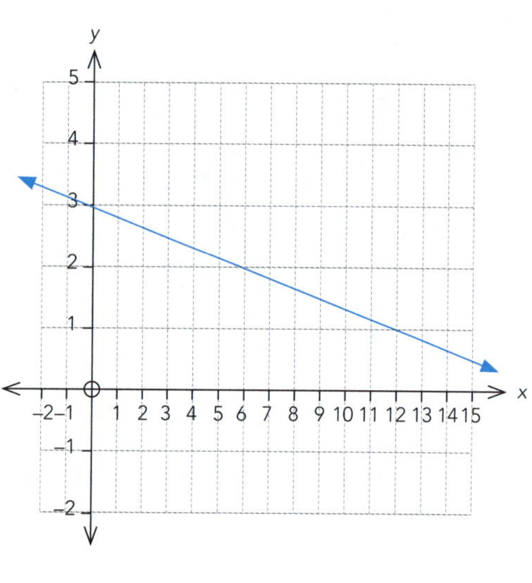

6

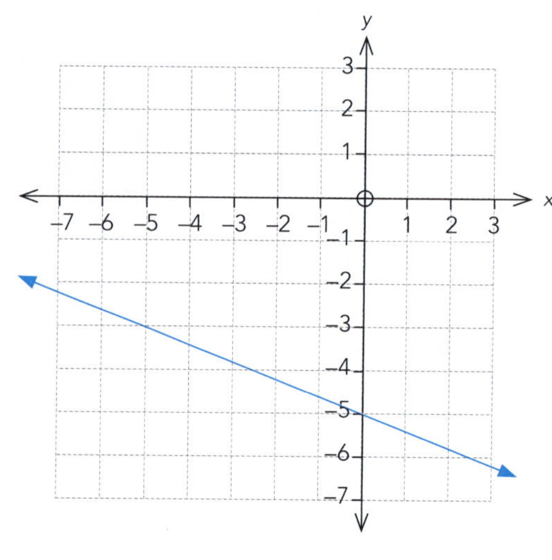

 ISBN: 9780170451468

7

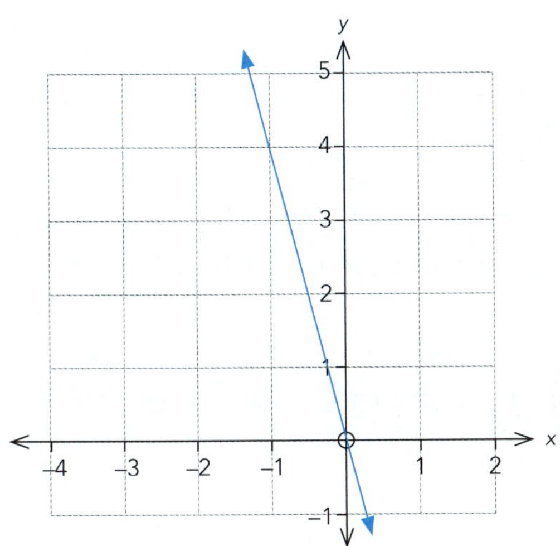

8

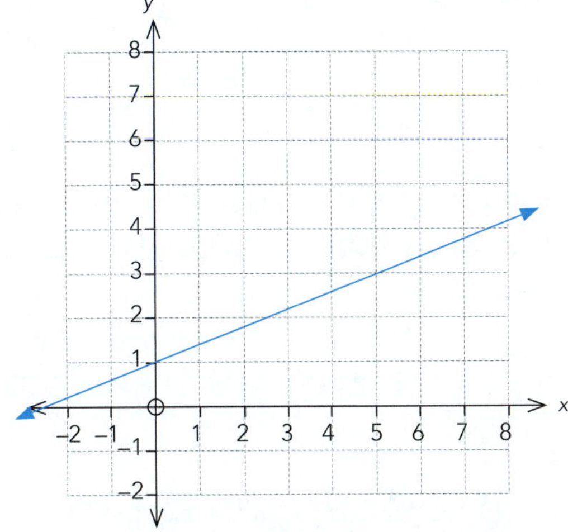

9

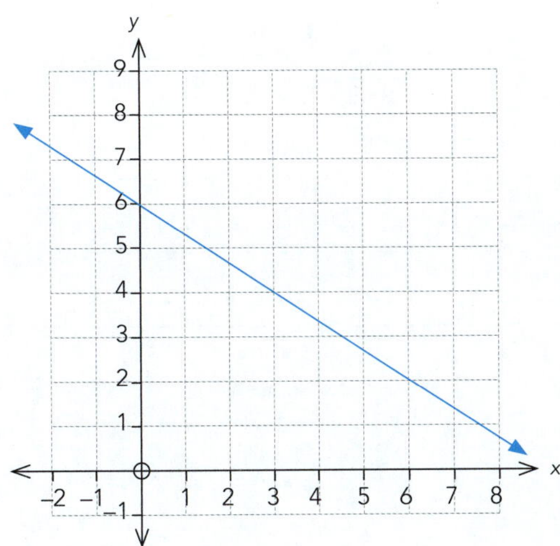

10

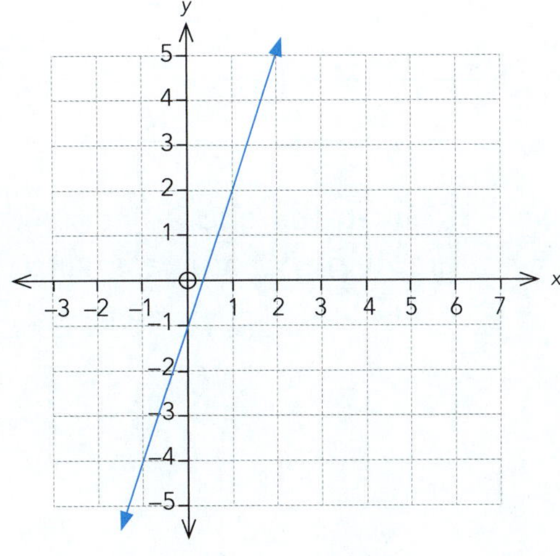

11

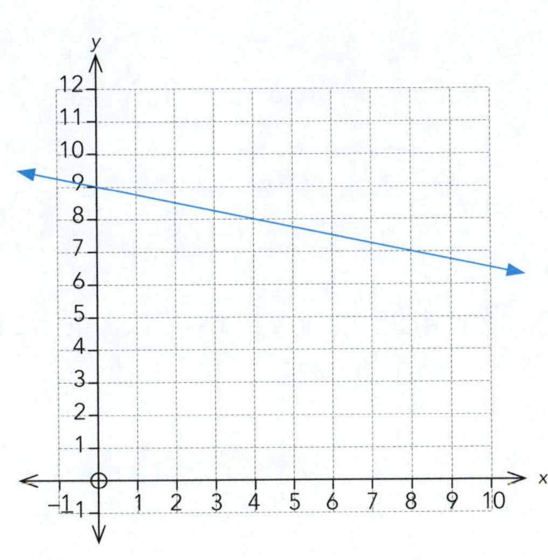

12

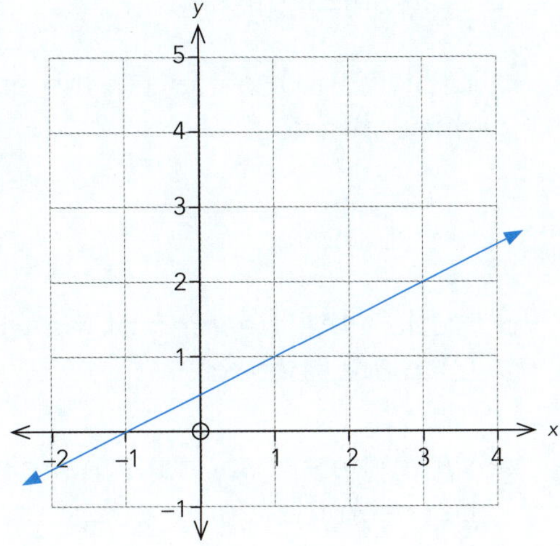

ISBN: 9780170451468

Applications

- Graphs of straight lines have many applications.

Example: Luca has a very short haircut to raise money for charity. The graphs shows the relationship between the number of weeks after the haircut and the length of his hair.

a Use the graph to help you complete the table to show the length of his hair as it grows.

Week (x)	Length (mm) (y)
0	1
2	6
4	11
6	16

Convert the grid points on the graph into coordinates and write them in the table.

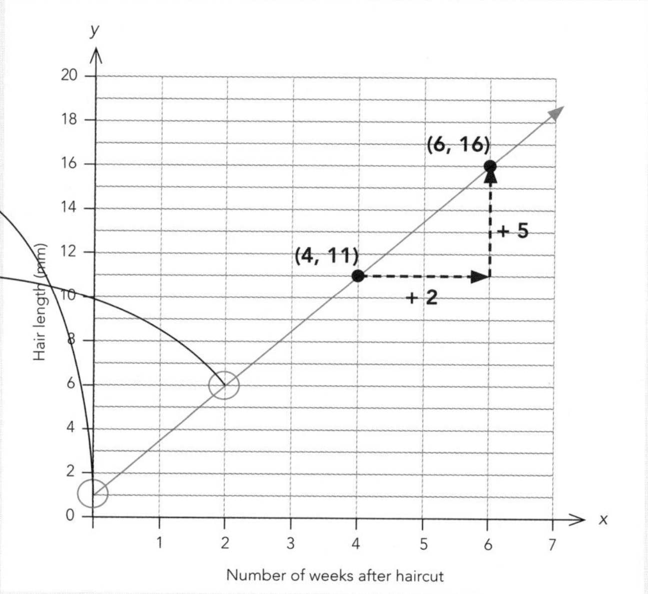

y
20
18
16
14
12
10
8
6
4
2
0
Hair length (mm)
(6, 16)
+ 5
(4, 11)
+ 2
1 2 3 4 5 6 7
x
Number of weeks after haircut

b Use the coordinates of the two points marked on the graph to calculate the gradient of the line.

$$m = \frac{5}{2}$$
$$= 2.5$$

c Explain the meaning of the gradient in this context.

Luca's hair grows 2.5 **mm** each **week**.

Always include **units** in your descriptions of gradients.

d Explain the meaning of the y intercept (c) in this context.

Immediately after the cut, his hair was 1 mm long.

e Write the equation for the relationship between the length of Luca's hair and the number of weeks since the haircut.

$y = 2.5x + 1$

f Use the equation to calculate the length of Luca's hair when at 20 weeks.

$y = 2.5 \times 20 + 1$
$= 51$ mm

ISBN: 9780170451468

Use the skills that you have learnt so far to answer these questions.

1 Jack enters a race. He runs the entire distance at a constant rate.

a Use the graph to help you complete the table to show how far he is from the finish line after each hour.

Time (h) (x)	Distance to the finish line (km) (y)
0	42
1	
2	
3	
4	
5	

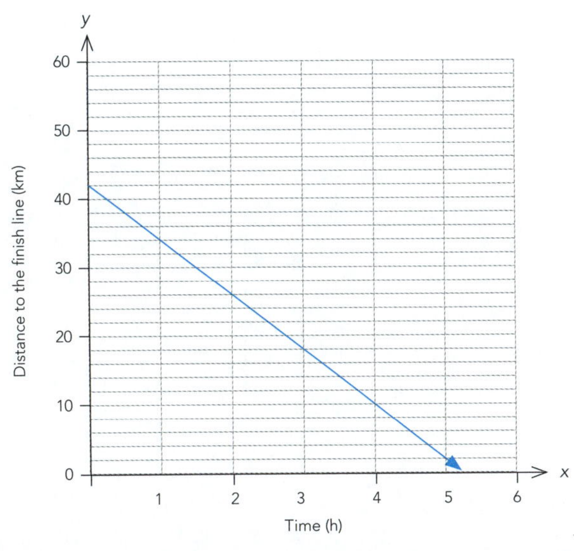

b Select two points on the graph and write the calculations to show that the gradient is –8.

Points: (_____, _____) and (_____, _____)

m = __________

= –8

c Explain what the gradient means in this context.

____

____

____

d The graph passes through the point (0, 42). Explain what this means in this context.

____

____

____

e Write an equation which relates Jack's distance to the finish line to the time he has been running.

$y =$ __________

f Use the equation to work out how far he is from the finish line after 5 hours.

____

____

____

g Use the line on the graph to **estimate** how long it took him to reach the finish line.

____

2 On New Year's Day, Stella's spa pool started leaking at a constant rate.

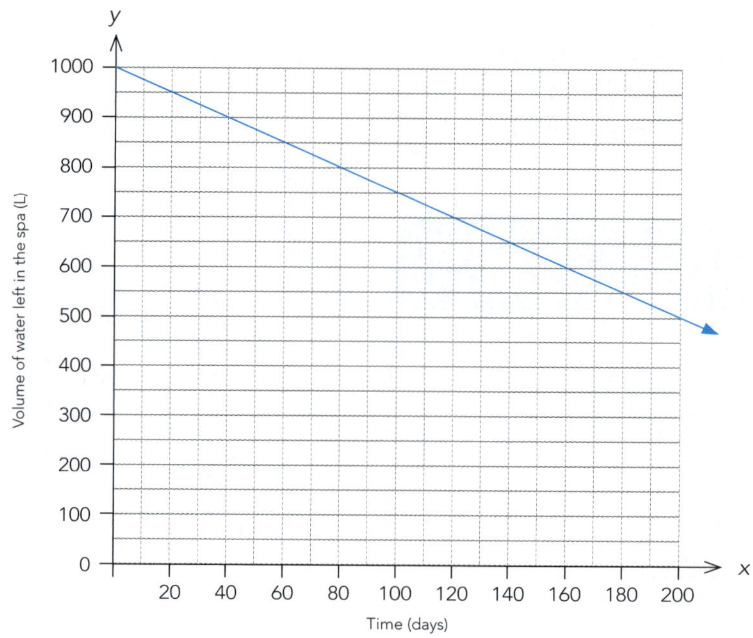

a Select two points on the graph and write the calculations to show that the gradient is –2.5.

Points: (______, ______) and (______, ______)

m = ______________

= –2.5

b Explain what the gradient means in this context.

____

____

____

c The graph passes through the point (0, 1000). Explain what this means in this context.

____

____

d Write an equation which relates the volume of water to the number of days since New Year's Day.

y = ______________

e Assume the leak was not repaired, and the spa pool was covered and not topped up. Use the equation to work out the volume of water remaining on New Year's Day the following year (365 days later).

____

____

____

ISBN: 9780170451468

3 There is a power cut, and Manu's freezer starts to warm up at a constant rate.

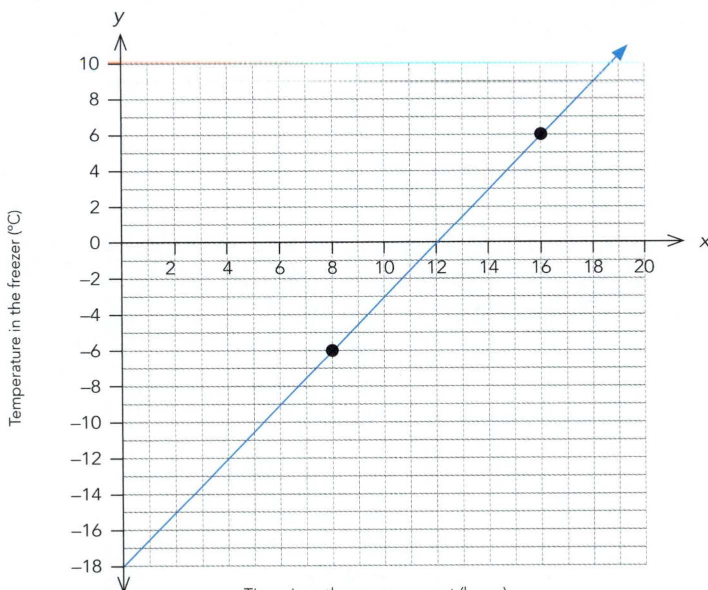

a Write the coordinates of the two points on the graph, and use these to find the gradient.

Points: (_____, _____) and (_____, _____)

m = __________

= __________

b Explain what the gradient means in this context.

c The graph passes through the point (0, −18). Explain what this means in this context.

d Write an equation which relates the temperature in the freezer to the number of hours since the power was cut.

y = __________

e Assume the power does not come back on, and the temperature continues to rise at the same rate. Use the equation to calculate the temperature after 14 hours.

f Describe how the line would change if the freezer warmed up more rapidly.

4 Forensic scientists have found that the relationship between the lengths of shoe imprints and the height of suspects is approximately linear for the range on the graph. The graph shows this relationship for males.

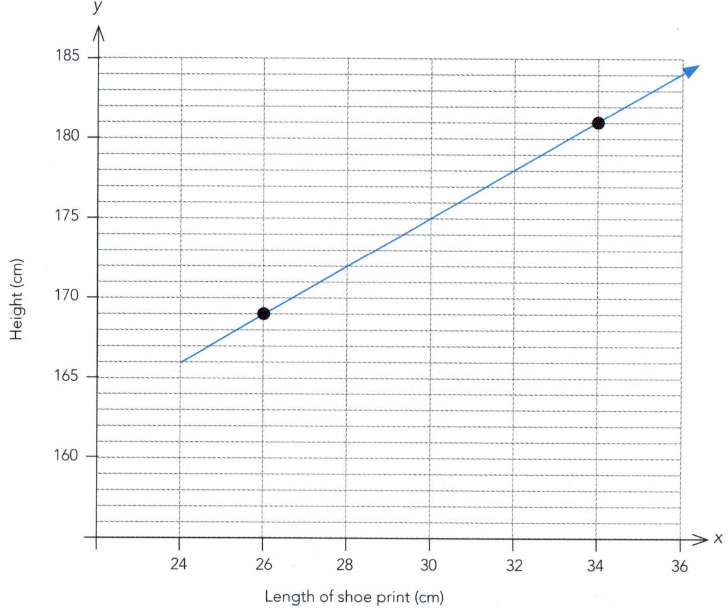

a Write the coordinates of the two points on the graph, and use these to find the gradient.

Points: (_____, _____) and (_____, _____)

m = ____________

= ____________

b Explain what the gradient means in this context.

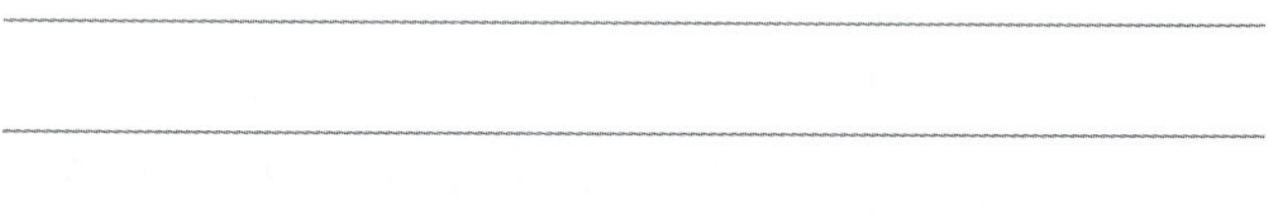

c The y intercept (not shown on the graph) is (0, 130). Write an equation which relates the height of a suspect to the length of a shoe print.

$y =$ ____________

d Use the equation to calculate the approximate height of a person whose shoe print is 31 cm long.

e Use the graph to **estimate** the shoe length of a male whose height is 1.8 m.

 ISBN: 9780170451468

5 The relationship between the mass of a chicken and the length of time it takes to cook at a particular temperature is approximately linear for chickens between 500 g and 3.5 kg. This is shown on the graph.

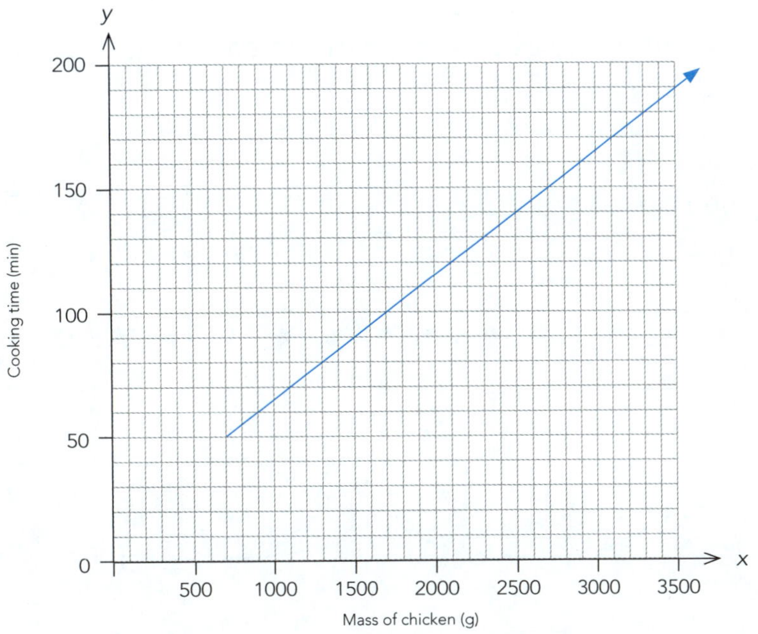

a Select two points on the graph and write the calculations to find the gradient.

Points: (____, ____) and (____, ____)

m = ________

= ________

b Explain what the gradient means in this context.

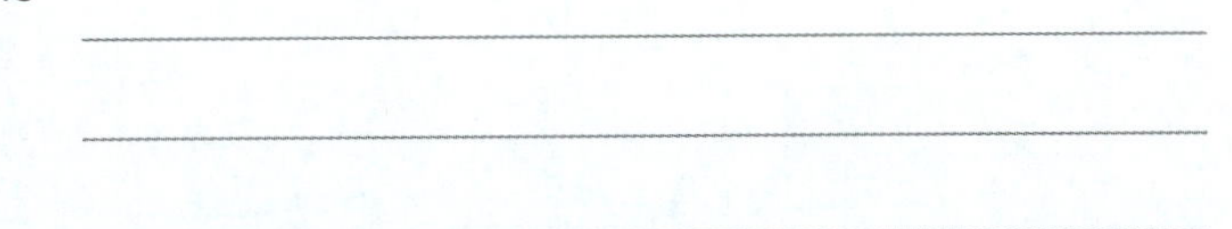

c The line could be extended through the point (0, 15). Write an equation which relates the cooking time of a chicken to its mass.

$y =$ ________

d Use the equation to calculate the cooking time for a chicken with mass 1.6 kg.

e If the oven temperature is increased, the graph becomes flatter. Explain what that means in this context.

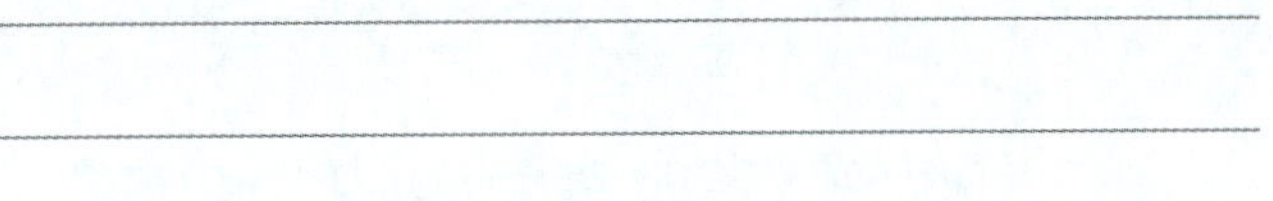

ISBN: 9780170451468

Linear or not?

- **Linear** patterns are patterns that increase or decrease by the **same amount** each time.
- A pattern that increases by **different amounts** each time is called a **non-linear** pattern.

Examples:

1 Consider the number of dots.

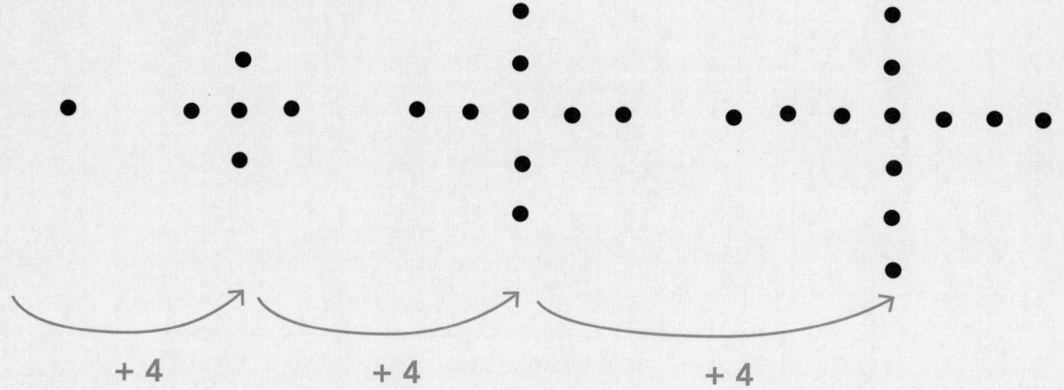

+ 4 + 4 + 4

This pattern increases by the **same number** (4) each time, therefore it **is** linear.

2 Consider the number of small squares.

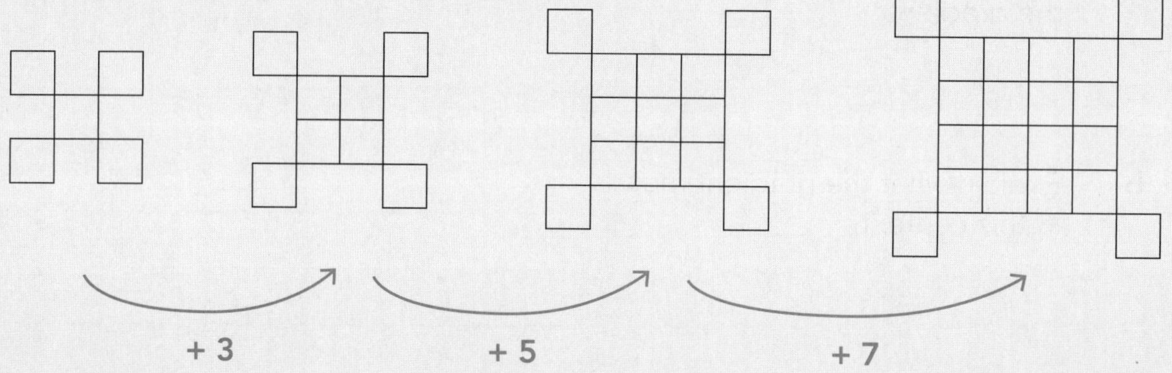

+ 3 + 5 + 7

This pattern increases by **different numbers** of small squares each time, therefore it is **not** linear.

3

45, 40, 35, 30, 25

− 5 − 5 − 5 − 5

This pattern decreases by the **same number** (5) each time, therefore it **is** linear.

4

1, 3, 7, 13, 21

+ 2 + 4 + 6 + 8

This pattern increases by **different numbers** each time, therefore it is **not** linear.

 ISBN: 9780170451468

Identify if these patterns are linear or non-linear.

1 Consider the number of popsicle sticks.

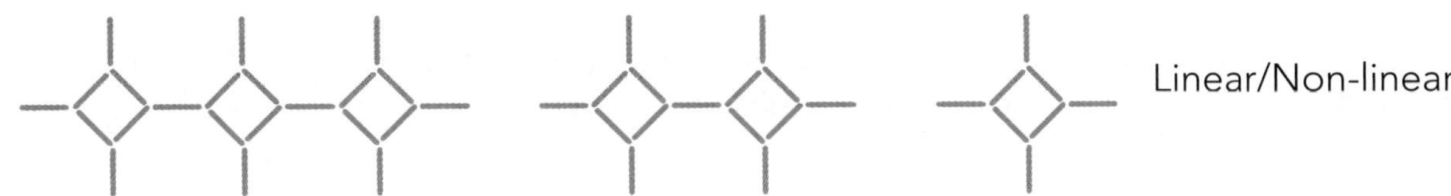

Linear/Non-linear

2 Consider the number of small squares.

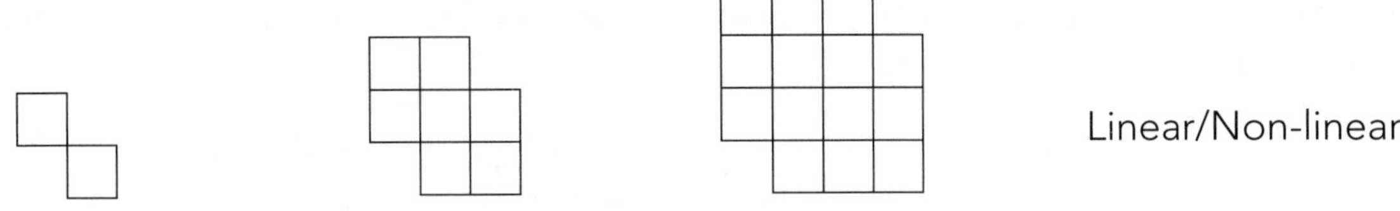

Linear/Non-linear

3 Consider the number of small triangles.

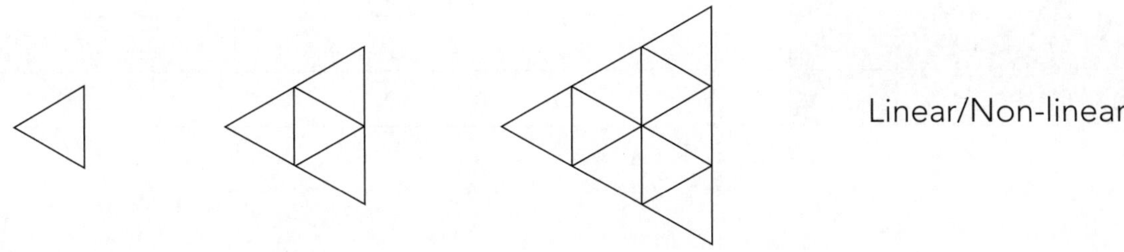

Linear/Non-linear

4 Consider the number of crosses.

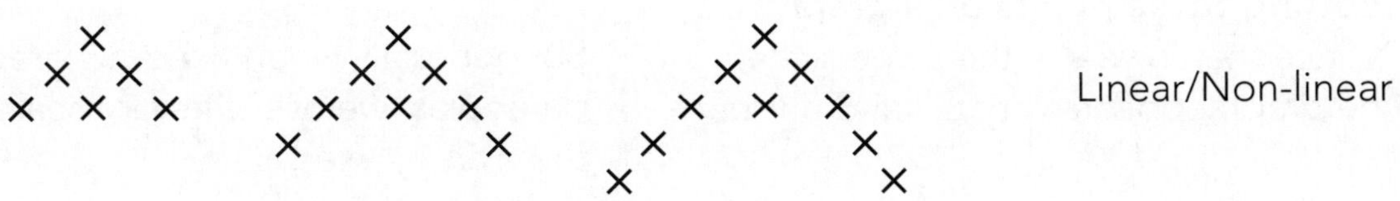

Linear/Non-linear

5 10, 7, 4, 1, –2, …

Linear/Non-linear

6 4, 10, 18, 28, 40, …

Linear/Non-linear

7 –4.5, –1.5, 1.5, 4.5, 7.5, …

Linear/Non-linear

ISBN: 9780170451468

Quadratic patterns

- A **quadratic equation** is an equation in which the **highest power of x is 2**.
- In this book we will consider quadratics in the following forms: $y = x^2$, $y = x^2 \pm b$, $y = -x^2$
- Usually quadratic equations contain an x^2, like those above, but we will also consider quadratic equations in factorised form which look like: $y = (x \pm c)^2$
- The **parabola** is the shape of the graph obtained when any **quadratic equation** is plotted.

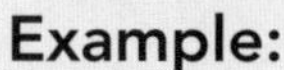

Example:

Term number (x)	1	2	3	4	5
	1^2	2^2	3^2	4^2	5^2
Number of dots (y)	1	4	9	16	25

$y = x^2$

+ 3 + 5 + 7 + 9

Notice that, unlike linear patterns, the increase from term to term is **not constant**.

Plotting these points on a graph:

Notice that if we use the same scale, the graph appears very tall and narrow.

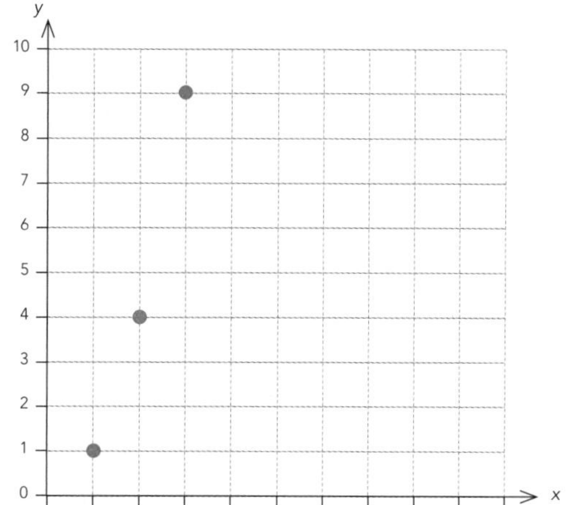

So, generally when we draw graphs of parabolas, we use different scales on the x and y axes.

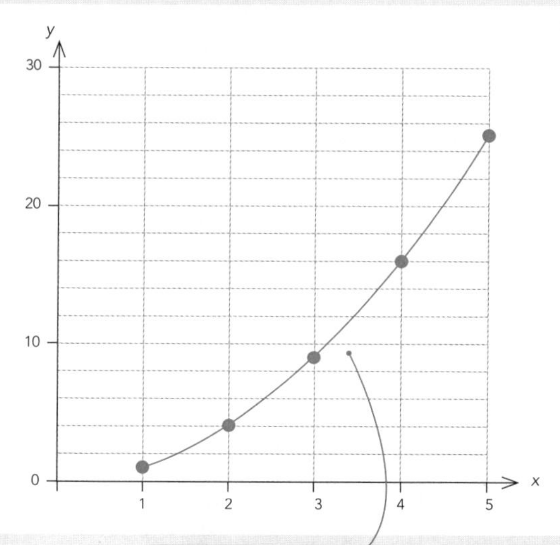

You need to join the points with a **smooth curve**.

 ISBN: 9780170451468

Extending the graph:
You can't just rule a longer line, as you can for linear graphs. Sometimes you need to **extend** the table and calculate more values:

There are extra points that you need to plot to get both sides of the parabola. (columns $x = -5$ to -1)

These are points you might plot for a straight line. (columns $x = 0$ to 5)

x	–5	–4	–3	–2	–1	0	1	2	3	4	5
x^2	$(-5)^2$	$(-4)^2$	$(-3)^2$	$(-2)^2$	$(-1)^2$	0^2	1^2	2^2	3^2	4^2	5^2
y	25	16	9	4	1	0	1	4	9	16	25

Plotting all these points produces a curve called a **parabola**. Its equation is $y = x^2$.

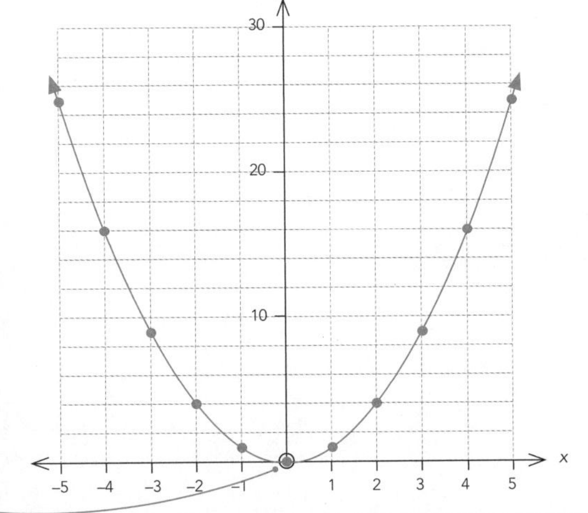

Notice: the base (or apex) of a parabola does **not** come to a point. It is a **smooth curve**.

Parabolas have a number of practical applications:

Trajectories of thrown objects.

Cross-sections of satellite dishes, headlight reflectors, heater reflectors, etc.

Calculations of stopping distances of cars, etc. in reaction to speed.

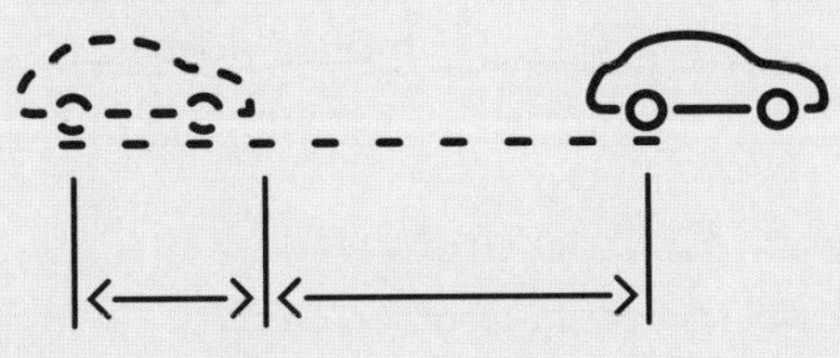

The areas of squares.

ISBN: 9780170451468

Drawing parabolas

1 In the form $y = x^2 \pm b$

Examples:

1 Draw the graph of $y = x^2 + 3$.

Step 1: Make a table.

x	$x^2 + 3$	y	Point
–3	$(-3)^2 + 3$	12	(–3, 12)
–2	$(-2)^2 + 3$	7	(–2, 7)
–1	$(-1)^2 + 3$	4	(–1, 4)
0	$0^2 + 3$	3	(0, 3)
1	$1^2 + 3$	4	(1, 4)
2	$2^2 + 3$	7	(2, 7)
3	$3^2 + 3$	12	(3, 12)

Step 2: Plot the points and join with a smooth curve.

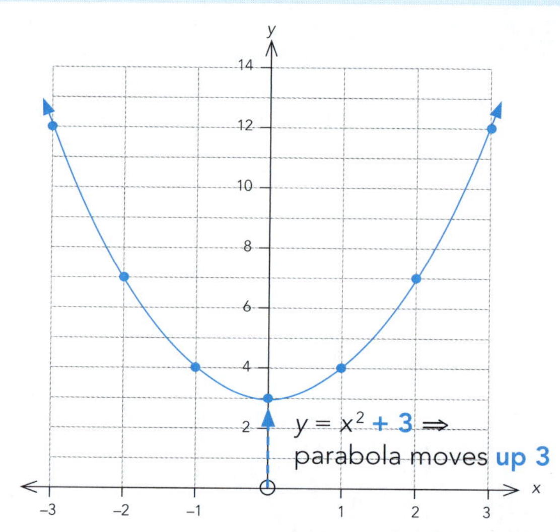

2 Draw the graph of $y = x^2 - 2$.

A physical pattern for this cannot be drawn because if $x = 1$, $y = -1$. However, it could be used as a model for other situations where negative values are possible.

Step 1: Make a table.

x	$x^2 - 2$	y	Point
–3	$(-3)^2 - 2$	7	(–3, 7)
–2	$(-2)^2 - 2$	2	(–2, 2)
–1	$(-1)^2 - 2$	–1	(–1, –1)
0	$0^2 - 2$	–2	(0, –2)
1	$1^2 - 2$	–1	(1, –1)
2	$2^2 - 2$	2	(2, 2)
3	$3^2 - 2$	7	(3, 7)

Step 2: Plot the points and join with a smooth curve.

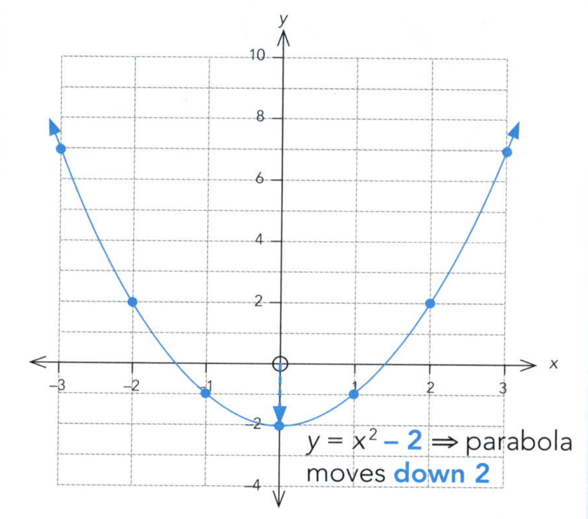

For parabolas of the form $y = x^2 \pm b$:

+ b ⇒ move b units **up**

– b ⇒ move b units **down**

 ISBN: 9780170451468

Complete the tables, plot the points, and join them to form a smooth curve.

1 $y = x^2 + 1$

x	$x^2 + 1$	y	Point
–3	$(-3)^2 + 1$	10	(–3, 10)
–2	$(-2)^2 + 1$		(–2, _____)
–1			(_____, _____)
0			(_____, _____)
1			(_____, _____)
2			(_____, _____)
3			(_____, _____)

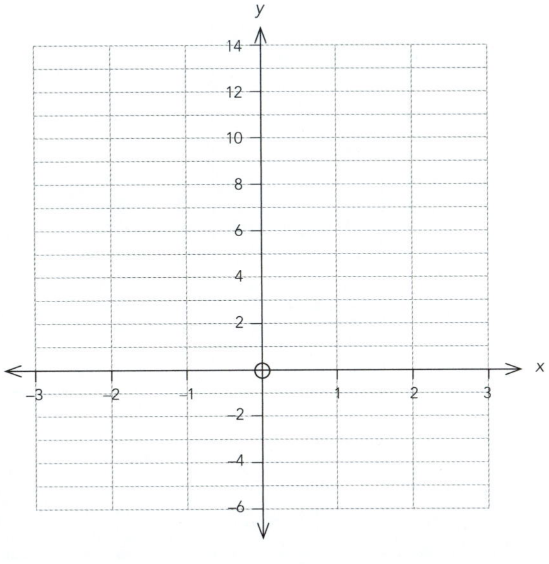

Compared with the graph of $y = x^2$, the parabola $y = x^2 + 1$ has moved up/down _____.

2 $y = x^2 + 5$

x	$x^2 + 5$	y	Point
–3	$(-3)^2 + 5$	14	(_____, _____)
–2			
–1			
0			
1			
2			
3			

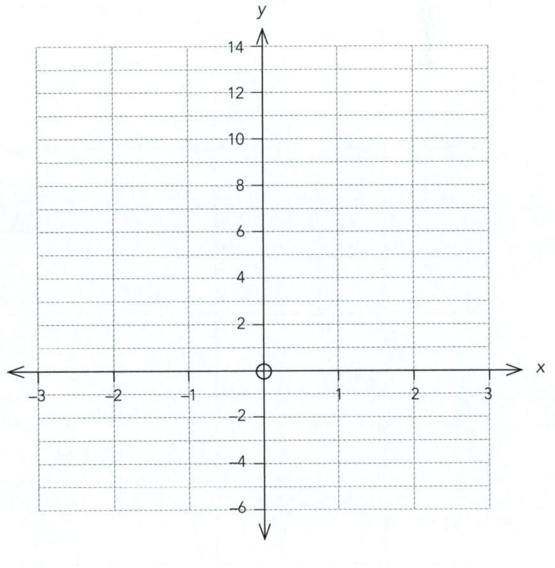

Compared with the graph of $y = x^2$, the parabola $y = x^2 + 5$ has moved up/down _____.

3 $y = x^2 - 4$

x	$x^2 - 4$	y	Point
–3	$(-3)^2 - 4$		
–2			
–1			
0			
1			
2			
3			

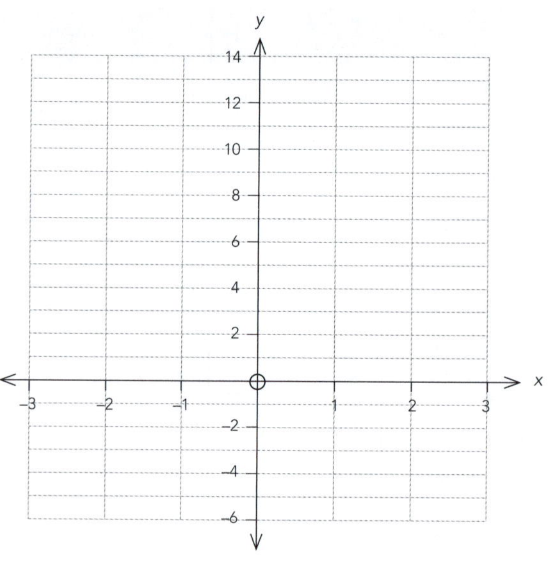

Compared with the graph of $y = x^2$, the parabola $y = x^2 - 4$ has moved up/down ______.

4 $y = x^2 - 6$

x	$x^2 - 6$	y	Point
–3			
–2			
–1			
0			
1			
2			
3			

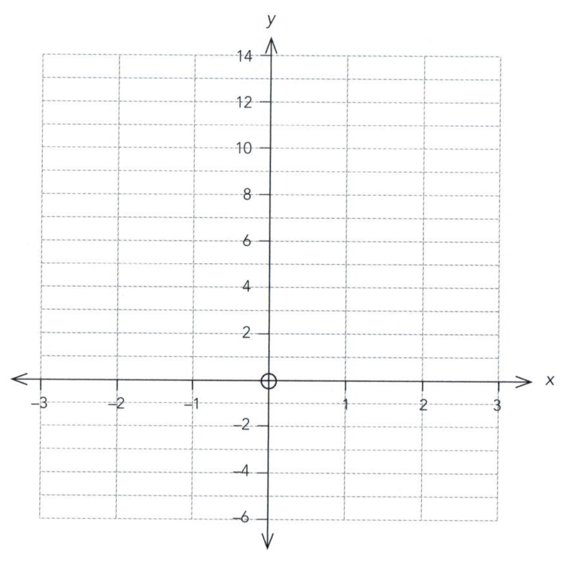

Compared with the graph of $y = x^2$, the parabola $y = x^2 - 6$ has moved up/down ______.

ISBN: 9780170451468

2 In the form $y = (x \pm c)^2$

Examples:

1 Draw the graph of $y = (x + 3)^2$.

Step 1: Make a table.

x	$(x + 3)^2$	y	Point
–3	$(-3 + 3)^2$	0	(–3, 0)
–2	$(-2 + 3)^2$	1	(–2, 1)
–1	$(-1 + 3)^2$	4	(–1, 4)
0	$(0 + 3)^2$	9	(0, 9)
1	$(1 + 3)^2$	16	(1, 16)
2	$(2 + 3)^2$	25	(2, 25)
3	$(3 + 3)^2$	36	(3, 36)
–4	$(-4 + 3)^2$	1	(–4, 1)
–5	$(-5 + 3)^2$	4	(–5, 4)
–6	$(-6 + 3)^2$	9	(–6, 9)

Notice that these points show only half the parabola.

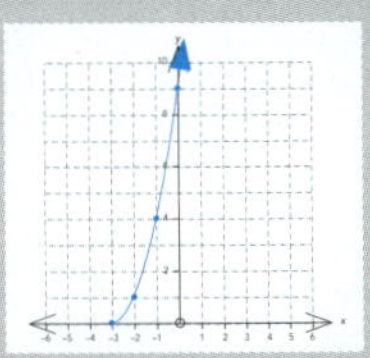

These points have very large values for y, so won't fit on the graph.

As a result, we need to add **more values of x** to the table.

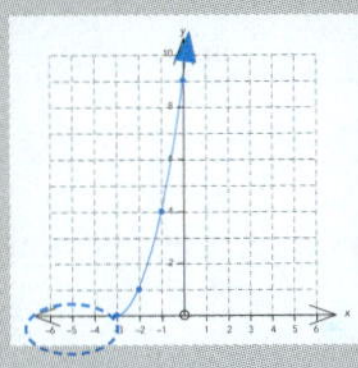

Step 2: Plot the points and join with a smooth curve.

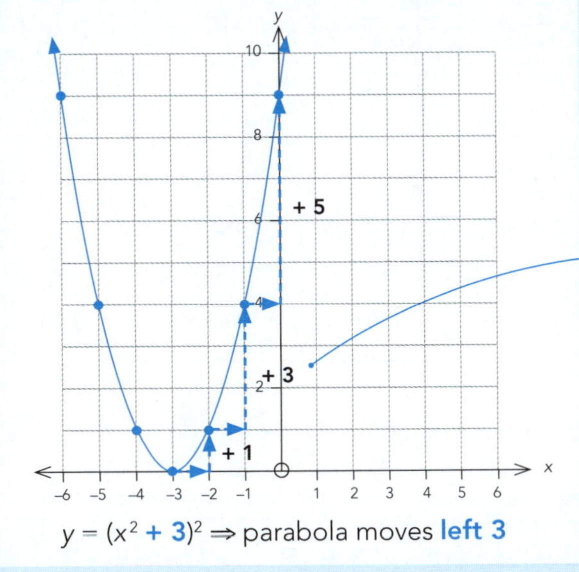

$y = (x^2 + 3)^2 \Rightarrow$ parabola moves **left 3**

Plotting parabolas of the form $y = x^2 \pm b$ or $y = (x \pm c)^2$ is easy if you remember that from the base, you go + 1, + 3, + 5 etc.

2 Draw the graph of $y = (x - 2)^2$.

Step 1: Make a table.

x	$(x - 2)^2$	y	Point
–3	(**–3** – 2)^2	25	(–3, 25)
–2	(**–2** – 2)^2	16	(–2, 16)
–1	(**–1** – 2)^2	9	(–1, 9)
0	(**0** – 2)^2	4	(0, 4)
1	(**1** – 2)^2	1	(1, 1)
2	(**2** – 2)^2	0	(2, 0)
3	(**3** – 2)^2	1	(3, 1)
4	(**4** – 2)^2	4	(4, 4)
5	(**5** – 2)^2	9	(5, 9)

These points have very large values for y, so won't fit on the graph.

Notice that these points show only part of the parabola.

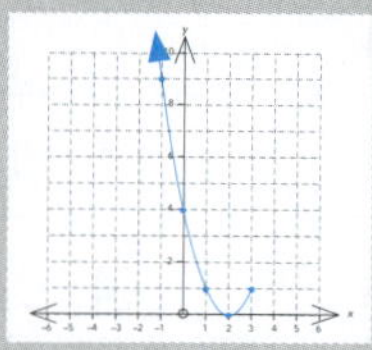

As a result, we need to add **more values of x** to the table.

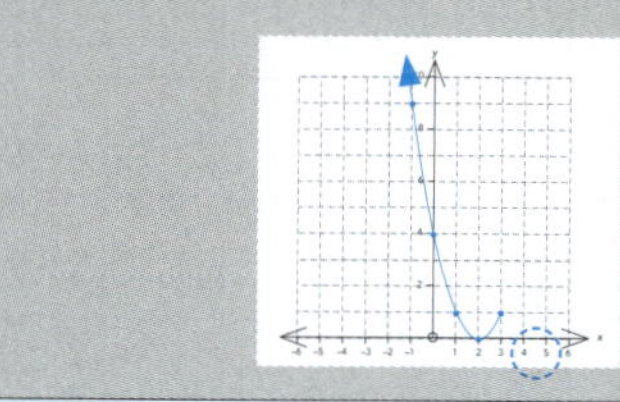

Step 2: Plot the points and join with a smooth curve.

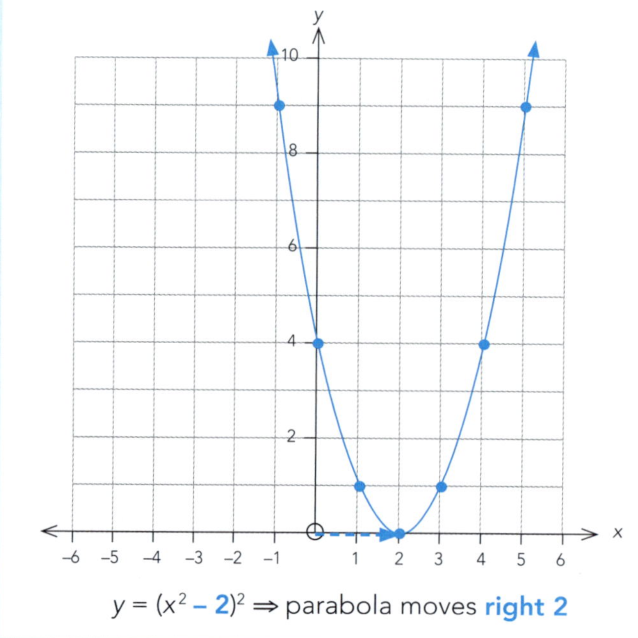

$y = (x^2 - \mathbf{2})^2 \Rightarrow$ parabola moves **right 2**

For parabolas of the form $y = (x \pm c)^2$:

+ c $\Rightarrow$ move c units **left**

– c $\Rightarrow$ move c units **right**

 ISBN: 9780170451468

Complete the tables, plot the points, and join them to form a smooth curve.

1 $y = (x + 1)^2$

x	$(x + 1)^2$	y	Point
–3	$(-3 + 1)^2$	4	(–3, 4)
–2	$(-2 + 1)^2$		(–2, ____)
–1			(____, ____)
0			(____, ____)
1			(____, ____)
2			(____, ____)
3			(____, ____)

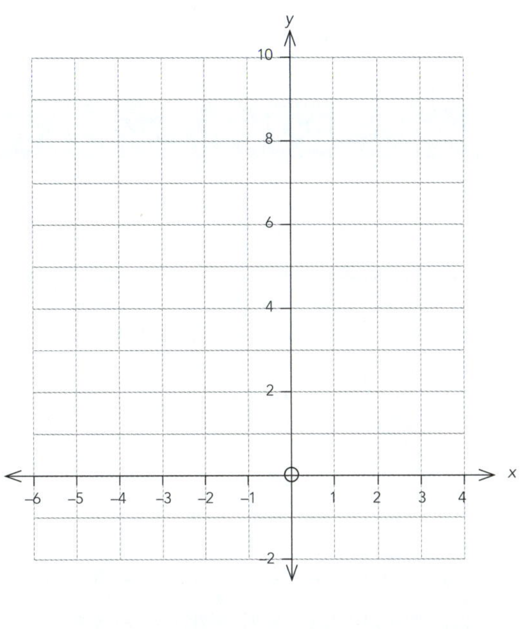

Spare line for other points.

Compared with the graph of $y = x^2$, the parabola $y = (x + 1)^2$ has moved right/left ____.

Hint: $(x + 4)^2$ tells you that $x = -4$ will be a useful point, so plot the points either side of it.

2 $y = (x + 4)^2$

x	$(x + 4)^2$	y	Point
–4			
–3			
–2			
–1			
0		16	
–5			
–6			
–7			

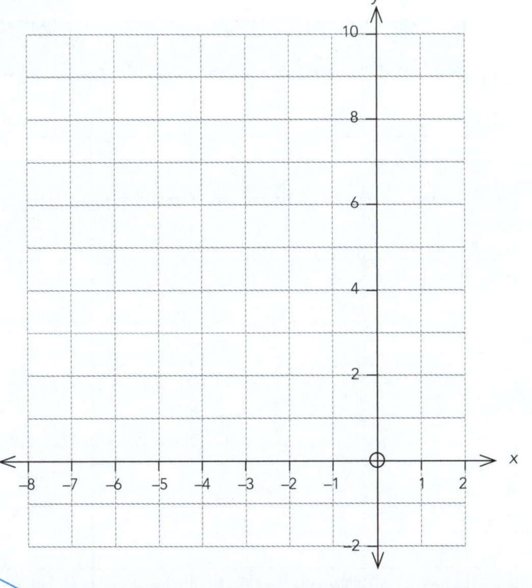

Too big, so plot the points you have, then look at your graph and decide what points you need.

Compared with the graph of $y = x^2$, the parabola $y = (x + 4)^2$ has moved right/left ____.

ISBN: 9780170451468

For the following functions, select appropriate values of x, calculate their corresponding y coordinates, plot the points and join them to form a smooth curve.

3 $y = (x - 3)^2$

x	$(x - 3)^2$	y	Point

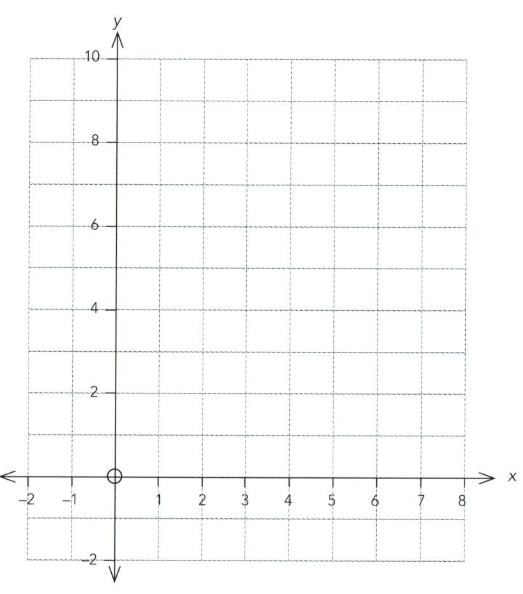

Compared with the graph of $y = x^2$, the parabola $y = (x - 3)^2$ has moved right/left ______.

4 $y = (x - 5)^2$

x	$(x - 5)^2$	y	Point

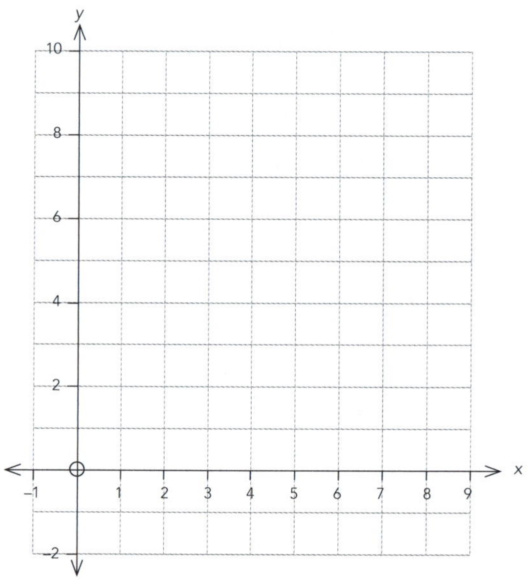

Compared with the graph of $y = x^2$, the parabola $y = (x - 5)^2$ has moved right/left ______.

 ISBN: 9780170451468

3 In the form $y = -x^2$

Complete the table, plot the points, and join them to form a smooth curve.

$y = -x^2$

x	$-x^2$	y	Point
–3	$-(-3)^2$	–9	(–3, –9)
–2			
–1			
0			
1			
2			
3			

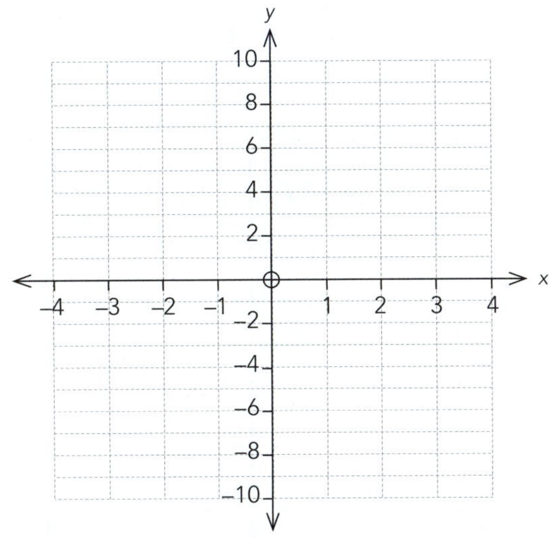

Compared with the graph of $y = x^2$, the parabola $y = -x^2$ is ______________________.

Putting it together

Select the correct equation for each graph.

$y = (x + 2)^2$ $y = x^2 + 2$ $y = (x - 2)^2$ $y = x^2 - 2$

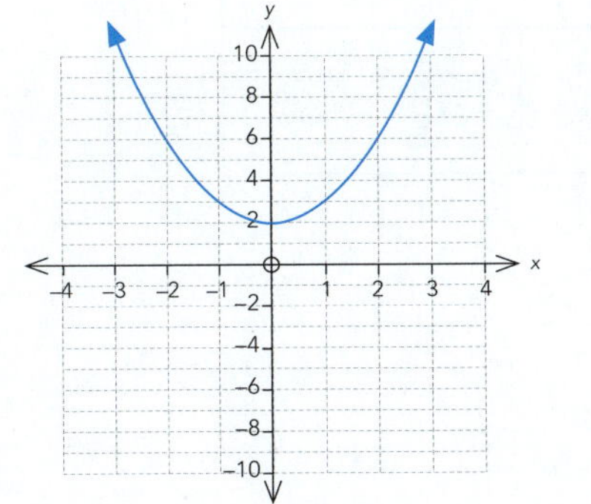

____

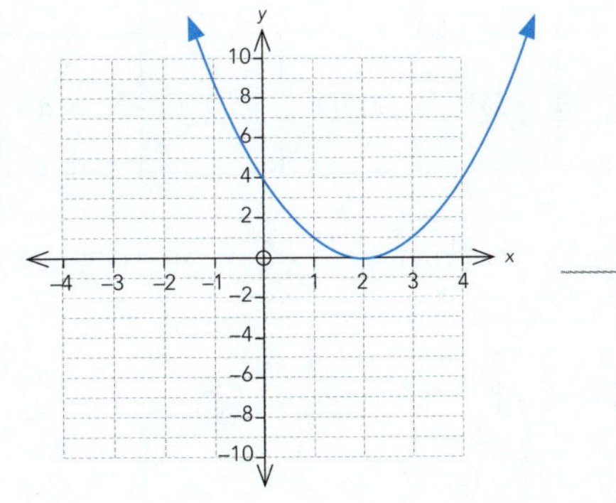

____

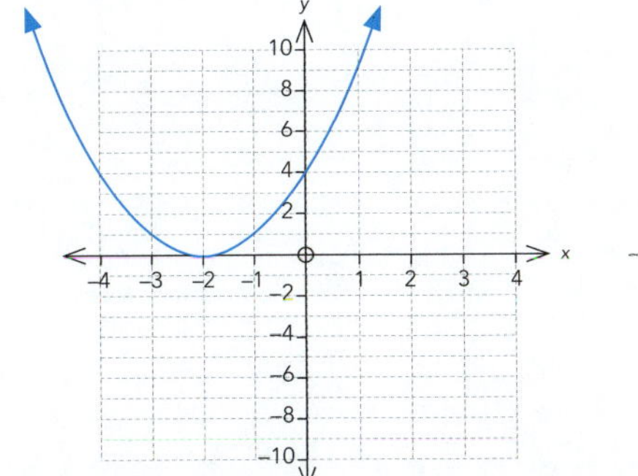

____

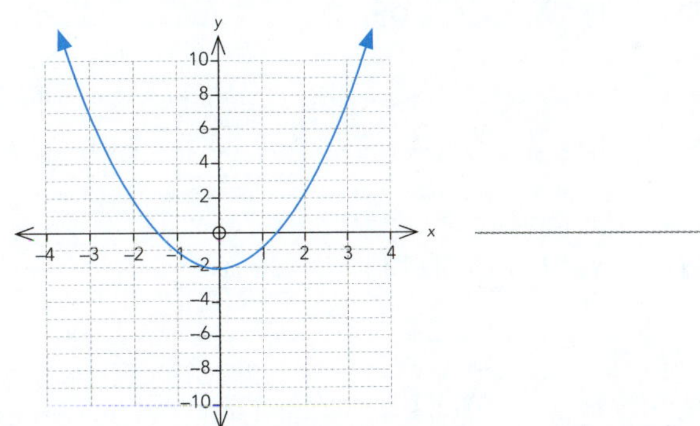

____

Challenge

1 Find and highlight the 13 sequences in this number grid. The sequences may be horizontal, vertical or diagonal, and each sequence is five terms long. Some may be different from those you have studied.

45	38	37	36	25	22	11	1	23	45
42	34	30	26	22	18	15	17	19	21
31	21	23	16	15	14	11	10	10	–3
21	17	14	11	8	5	1	4	1	4
41	23	15	7	–1	–9	–17	–10	–3	9
32	22	12	2	12	14	15	16	17	16
23	21	–11	–3	4	10	15	20	30	25
14	6	–18	–8	0	8	16	24	40	36
5	–1	–7	–13	–19	6	16	28	38	48
–3	–8	–20	–32	–44	–56	–66	32	34	60

2 Fill in the gaps so that you complete these patterns.

a 20, 19, 21, 18, 22, 17, ________, ________, ________

b 1, 8, 27, 64, ________, ________, ________

c 1, 1, 2, 3, 5, 8, 13, 21, ________, ________, ________

d o, t, t, f, f, ________, ________, ________

e m, t, w, t, ________, ________, ________

 ISBN: 9780170451468

Revision 1

1 **a** Complete the table.

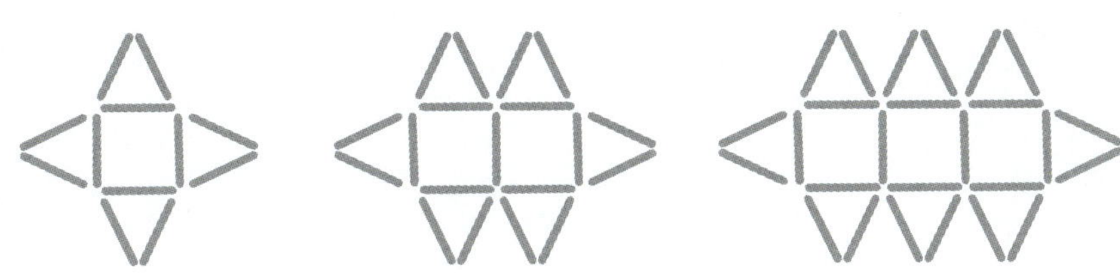

Shape number (n)	Number of popsicle sticks (T)
0	
1	
2	
3	
4	
5	
6	

b Find the rule.

Number of popsicle sticks = ________ x shape number + ________

T = ________________________

c How many popsicle sticks would be needed for the 50th shape?

T = ________________________

= ________________________

2 Complete the table and find the rule.
Start at 98 and reduce by 3 each time.

Term number (n)	0	1	2	3	4	5
Value (T)						

Rule:

T = ______n + ______

3 Find the rules for these sequences.

a 13, 18, 23, 28, 33, … Rule ________________ The 25th term = ________________

b 56, 45, 34, 23, 12, … Rule ________________ The 19th term = ________________

4 Use the rule $T = 5n - 3$ to calculate the first five terms of the sequence.

Term number (n)	1	2	3	4	5
Calculations					
Value (T)					

ISBN: 9780170451468

5 **a** Huia has 28 mints left in a jar. She eats 3 each day. Complete the table and graph to show the number of mints left in the jar after x days.

Days (x)	Mints (y)
0	
1	
2	
3	
4	
5	

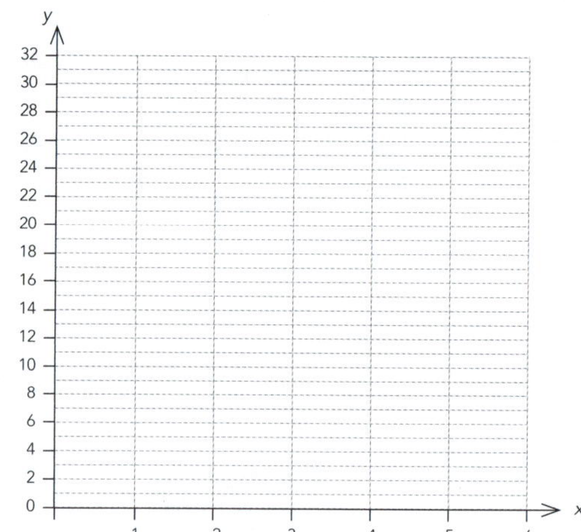

b Write the equation: $y =$ ________

c How many mints will Huia have after 9 days? ________

6 Use the equation to complete the table, plot the points on the graph, and then draw the line which represents $y = 4x - 1$.

x	y	Coordinates
0		
1		
2		
3		
4		
5		
6		

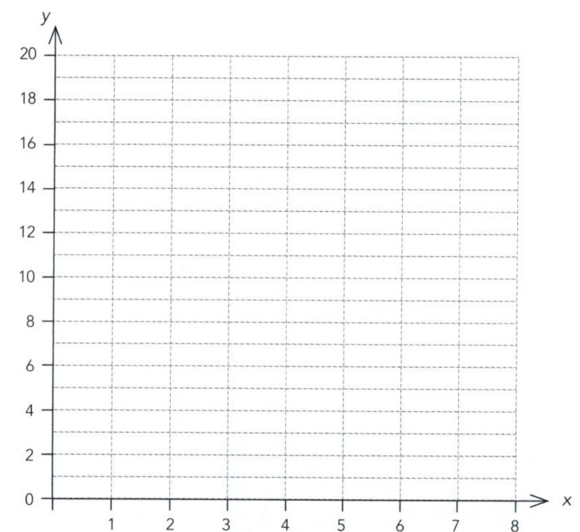

7 Use the graph to complete the table and find the equation.

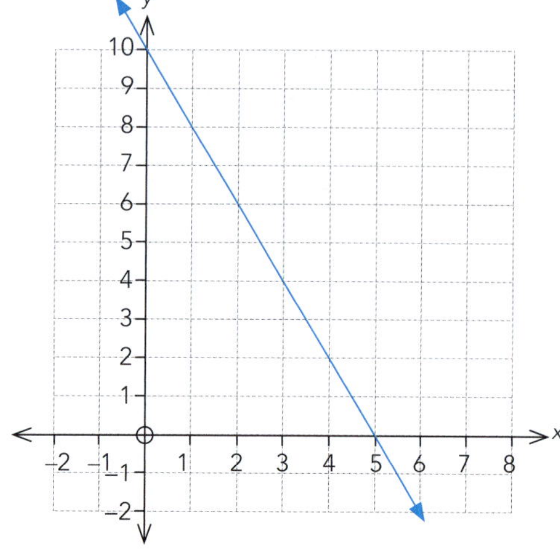

x	y
0	
1	
2	
3	
4	

Equation: $y =$ ________

 ISBN: 9780170451468

8 **a** Calculate the gradient of the line.

$m =$ ____________

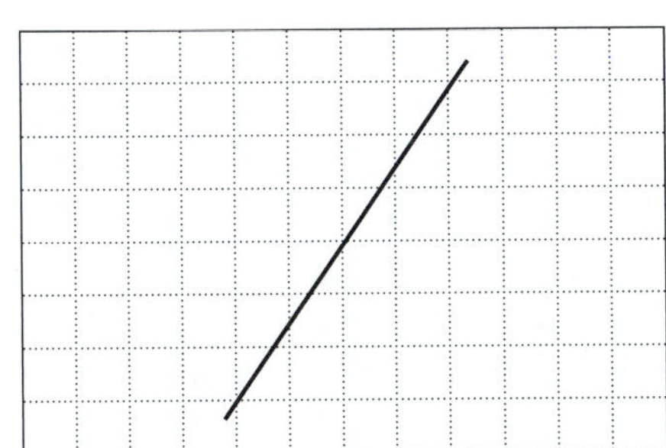

b Draw a line segment with a gradient of $m = -\frac{5}{6}$.

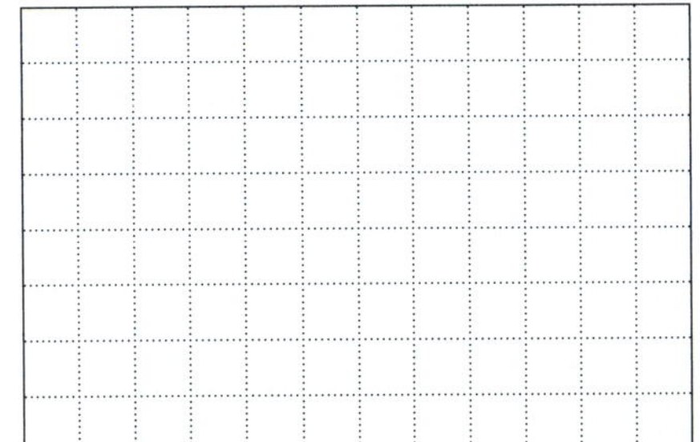

9 **a** Draw the graph of $y = \frac{4}{3}x - 2$.

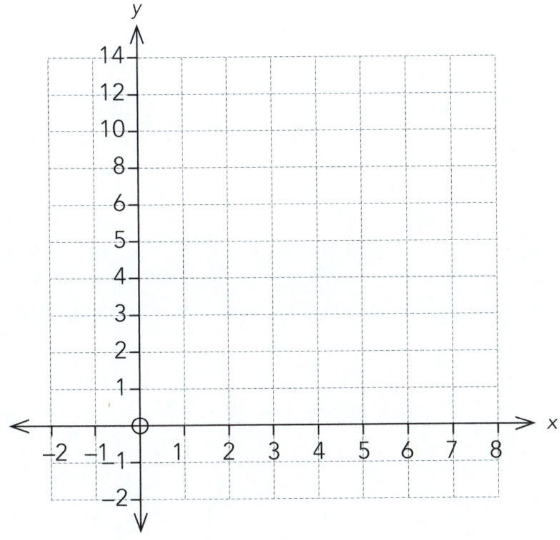

b Write the equation of the graph.

____

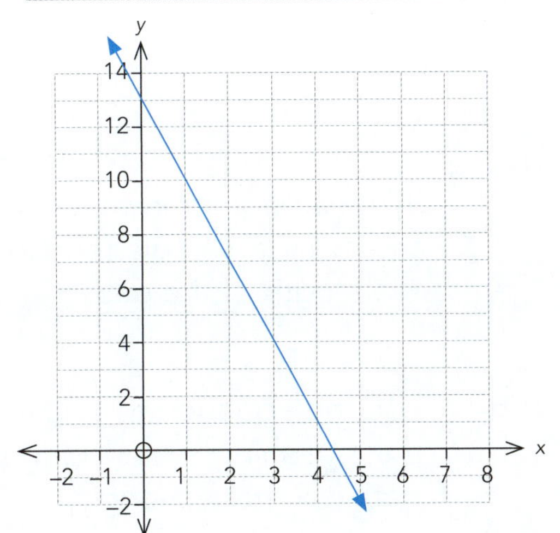

10 Complete the table, plot the points, and join them to form a smooth curve.
$y = x^2 + 2$

x	$x^2 + 2$	y	Point
–3	$(-3)^2 + 2$		
–2			
–1			
0			
1			
2			
3			

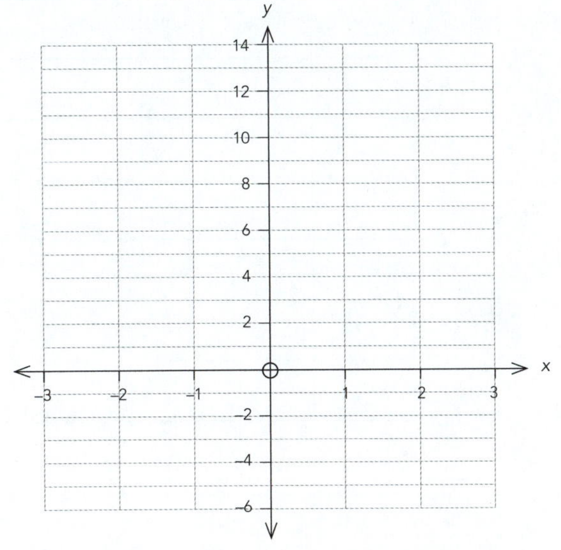

Compared with the graph of $y = x^2$, the parabola $y = x^2 + 2$ has moved up/down ______.

ISBN: 9780170451468

11 Nigel is building a path which is to be covered with shingle. Each metre of path needs the same mass of shingle, and he has had one tonne of shingle delivered. The graphs shows the relationship between the length of his path and the mass of shingle remaining.

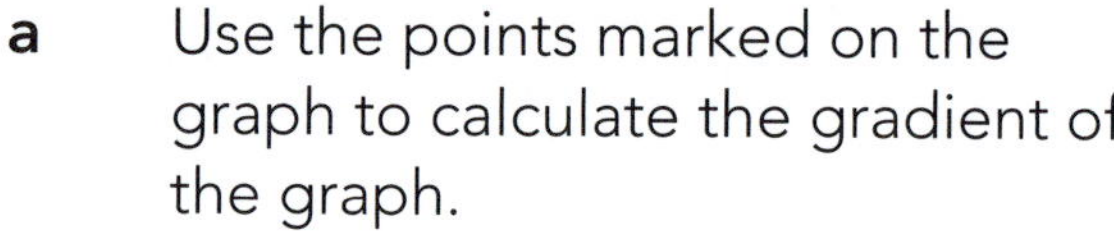

a Use the points marked on the graph to calculate the gradient of the graph.

____

____

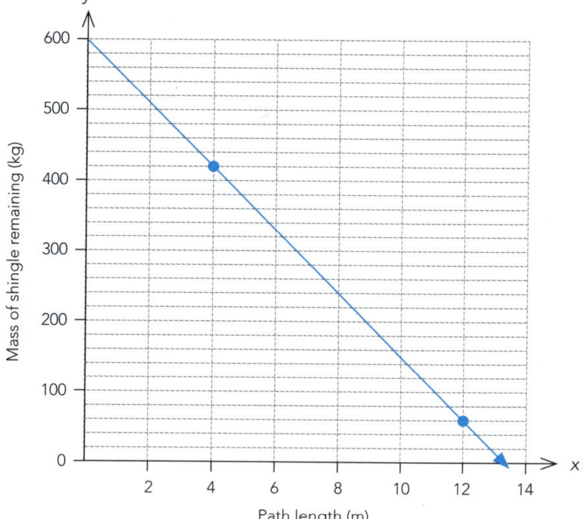

b Explain what the gradient means in this context.

____

____

c Write an equation which relates the mass of shingle remaining to the length of the path.

$y =$ ______________________________

d Use the equation to calculate the mass of shingle remaining if the path is 9 metres long.

____

____

12 **a** Complete the table and draw the graph of $y = (x - 2)^2$.

x	$(x - 2)^2$	y	Point
–3			
–2			
–1			
0			
1			
2			
3			

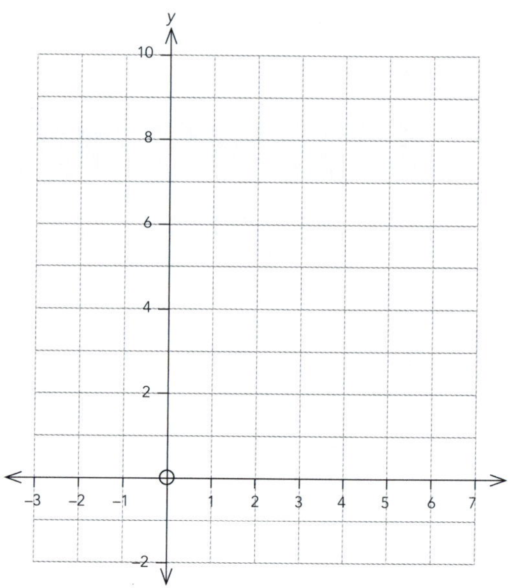

b Compared with the graph of $y = x^2$, the parabola $y = (x - 2)^2$ has moved right/left ______.

 ISBN: 9780170451468

Revision 2

1 **a** Complete the table.

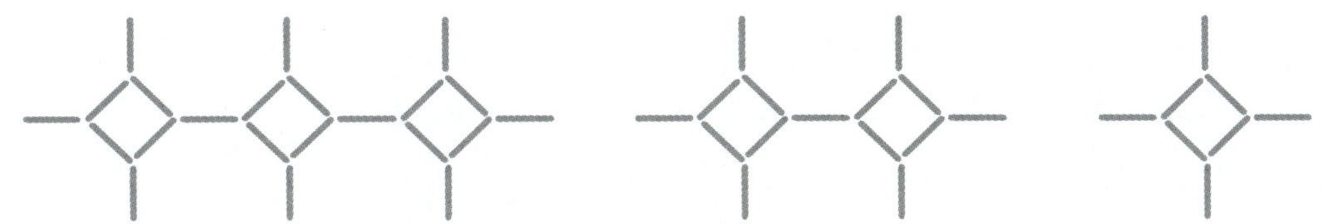

Shape number (n)	0	1	2	3	4
Number of popsicle sticks (T)					

b Find the rule.

Number of popsicle sticks = ______ x shape number + ______

T = ____________

2 Complete the table and find the rule.
Decreasing multiples of 7 that are smaller than 85.

Term number (n)	0	1	2	3	4	5
Value (T)						

Rule:

T = ______n + ______

3 Find the rules for these sequences.

a –2, 6, 14, 22, 30, … Rule ____________ The 36th term = ____________

b 37, 28, 19, 10, 1, … Rule ____________ The 15th term = ____________

4 Use the rule $T = -2n + 6$ to calculate the first five terms of the sequence.

Term number (n)	1	2	3	4	5
Calculations					
Value (T)					

ISBN: 9780170451468

5 **a** Hone is knitting himself a beanie. His first row has 6 stitches and he adds 2 stitches for each following row. Complete the table and graph to show the number of stitches after x rows.

Rows (x)	Stitches (y)
0	
1	
2	
3	
4	
5	

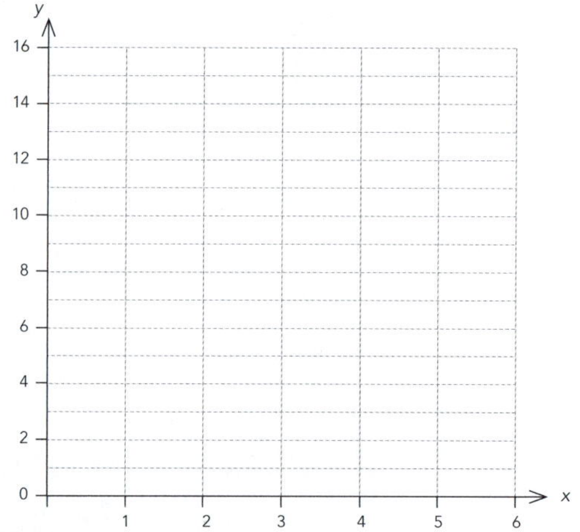

b Write the equation: $y =$ ________

c How many stitches will Hone have after 8 rows? ____________

6 Use the equation to complete the table, plot the points on the graph, and then draw the line which represents $y = -2x + 16$.

x	y	Coordinates
0		
1		
2		
3		
4		
5		
6		

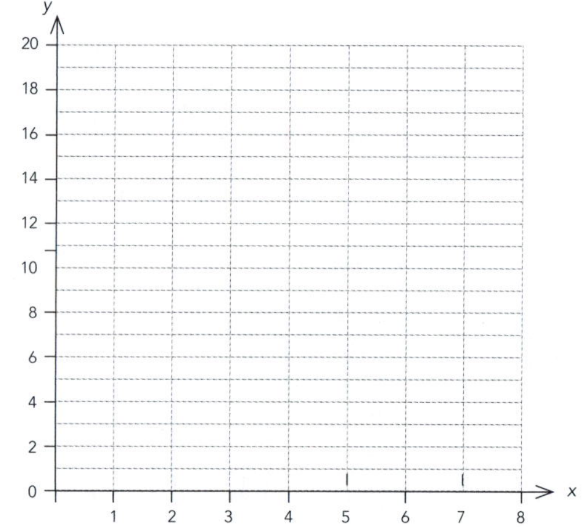

7 Use the graph to complete the table and find the equation.

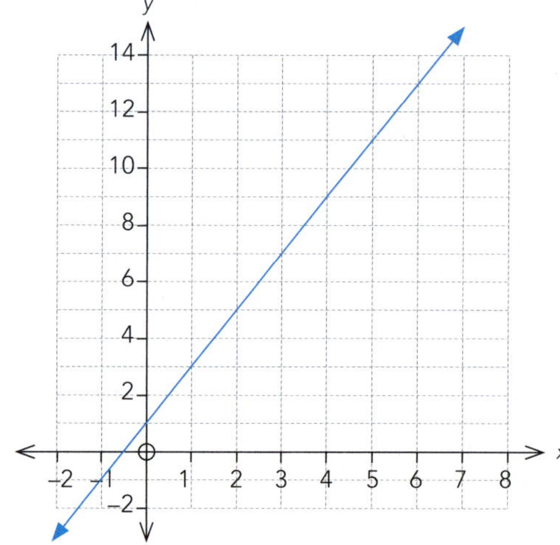

x	y
0	
1	
2	
3	
4	

Equation: $y =$ ________

 ISBN: 9780170451468

8 **a** Calculate the gradient of the line.

m = 

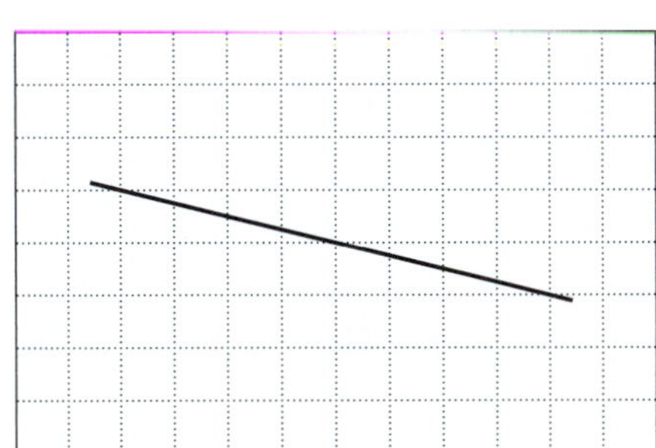

b Draw a line segment with a gradient of $m = \frac{4}{3}$.

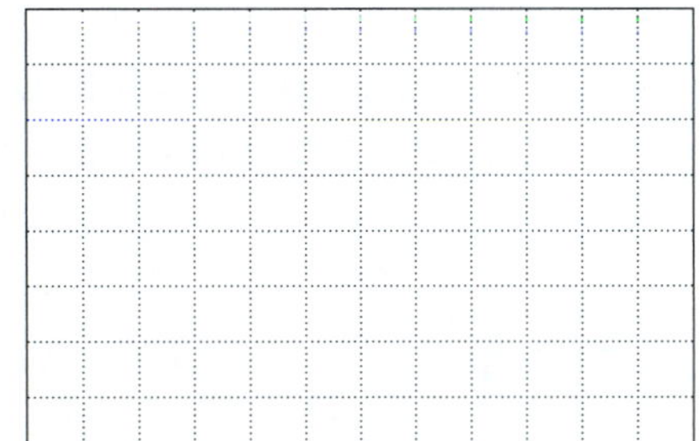

9 **a** Draw the graph of $y = -\frac{1}{2}x + 14$.

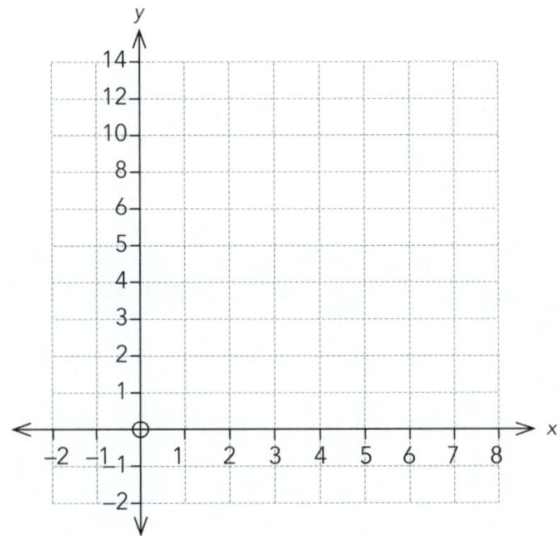

b Write the equation of the graph.

____

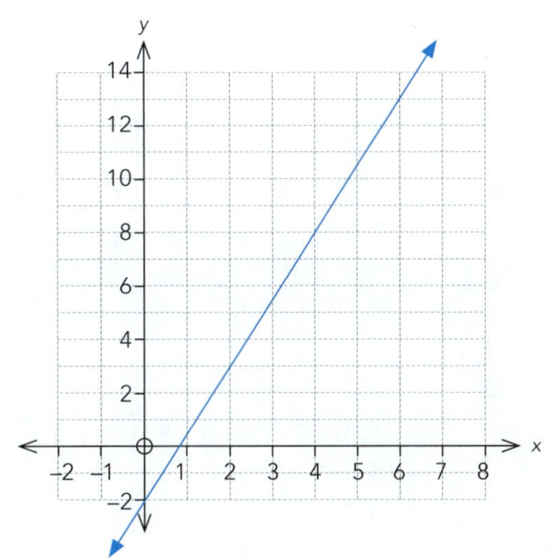

10 Complete the table, plot the points, and join them to form a smooth curve.
$y = x^2 - 5$

x	$x^2 - 5$	y	Point
–3	$(-3)^2 - 5$		
–2			
–1			
0			
1			
2			
3			

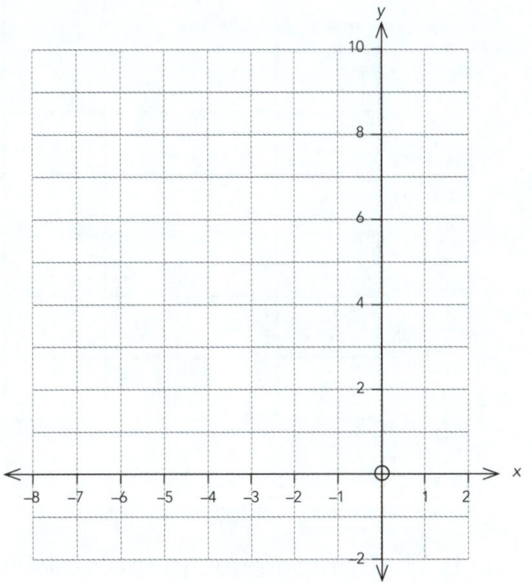

Compared with the graph of $y = x^2$, the parabola $y = x^2 - 5$ has moved up/down ______.

11 Hone is also knitting himself a scarf. He has found that for every centimetre he adds to its length, he needs x metres of wool. The graphs shows the relationship between the length of the scarf and the length of wool required.

a Use the points marked on the graph to calculate the gradient of the graph.

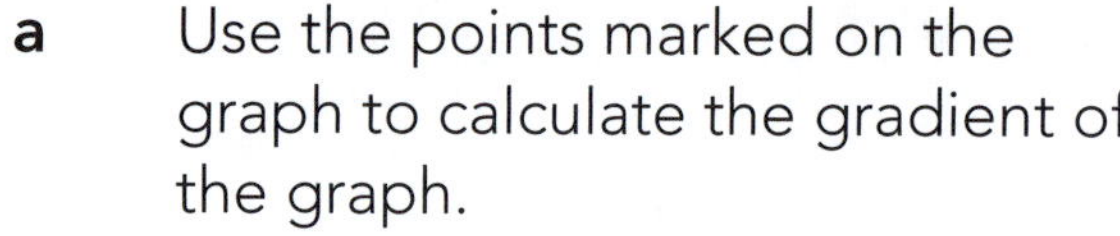

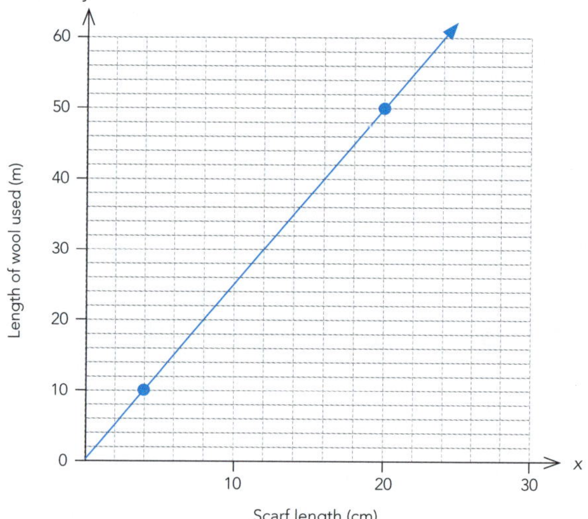

b Explain what the gradient means in this context.

c Write an equation which relates the length of wool used to the length of the scarf.

$y =$ ____________________

d Use the equation to calculate the length of wool used in a 1.2 m scarf.

12 **a** Complete the table and draw the graph of $y = (x + 3)^2$.

x	$(x + 3)^2$	y	Point
–3			
–2			
–1			
0			
1			
2			
3			

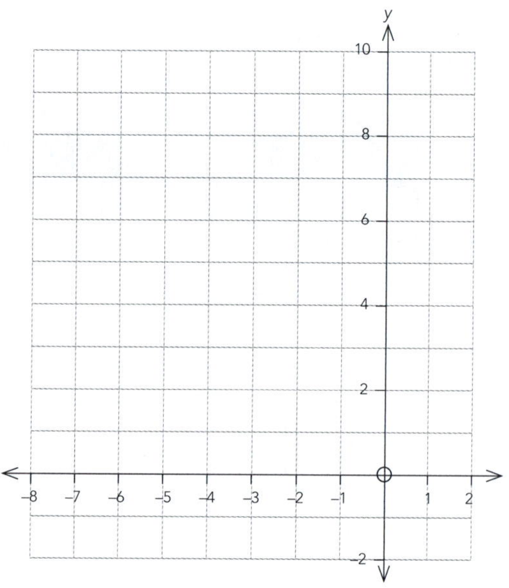

b Compared with the graph of $y = x^2$, the parabola $y = (x + 3)^2$ has moved right/left ______.

ISBN: 9780170451468

Answers

Revision: substitution (p. 5)

Formula	$n = 3$	$n = 6$	$n = -2$
$A = n + 4$	$A = 7$	$A = 10$	$A = 2$
$A = 6n$	$A = 18$	$A = 36$	$A = -12$
$A = 2n + 3$	$A = 9$	$A = 15$	$A = -1$
$A = 3n - 5$	$A = 4$	$A = 13$	$A = -11$
$A = -2n - 4$	$A = -10$	$A = -16$	$A = 0$

Linear patterns (pp. 6–28)

Patterns into tables (pp. 7–8)

1

Shape number	Number of dots
1	5
2	8
3	11
4	14
5	17
6	20

+ 3 + 3 + 3 + 3 + 3

2

Shape number	1	2	3	4	5	6
Number of popsicle sticks	7	11	15	19	23	27

+ 4 + 4 + 4

3

Shape number	Number of crosses
1	15
2	12
3	9
4	6
5	3
6	0

− 3

4

Shape number	1	2	3	4	5	6
Number of diamonds	16	14	12	10	8	6

−2

Finding term 'zero' (pp. 9–10)

1

Term number	0	1	2	3	4	5	6
Number of diamonds	1	5	9	13	17	21	25

2

Term number	Number of popsicle sticks
0	13
1	11
2	9
3	7
4	5
5	3

3

Term number	0	1	2	3	4	5	6
Number of dots	5	8	11	14	17	20	23

4

Term number	Number of popsicle sticks
0	28
1	23
2	18
3	13
4	8
5	3

ISBN: 9780170451468

Finding the rule from a pattern (pp. 11–18)

1

Term number (n)	Number of diamonds (D)
0	4
1	9
2	14
3	19
4	24
5	29
6	34

In words: I started with 9 diamonds and then I added/~~subtracted~~ 5 each time.
Find the mathematical rule:
Number of diamonds = 5 x term number + 4
Tidy it up: $D = 5n + 4$
Word rule: The number of diamonds is calculated by multiplying the term number by 5 and then adding/~~subtracting~~ 4.
Use the rule: $D = 5 \times 50 + 4$
$= 254$

2

Term number (n)	0	1	2	3	4	5	6
Number of squares (S)	12	11	10	9	8	7	6

In words: I started with 11 squares and then I ~~added~~/subtracted 1 each time.
Find the mathematical rule:
Number of squares = –1 x term number + 12
Tidy it up: $D = -1n + 12$
Word rule: The number of squares is calculated by multiplying the term number by –1 and then adding/~~subtracting~~ 12.
Use the rule: $S = -1 \times 10 + 12$
$= 2$

3

Term number (n)	Number of dots (D)
0	2
1	7
2	11
3	15
4	19
5	23
6	27

In words: I started with 7 dots and then I added/~~subtracted~~ 4 each time.
Find the mathematical rule:
Number of dots = 4 x term number + 3
Tidy it up: $D = 4n + 3$
Word rule: The number of dots is calculated by multiplying the term number by 4 and then adding/~~subtracting~~ 3.
Use the rule: $D = 4 \times 40 + 3$
$= 163$

4

Term number (n)	Number of popsicle sticks (P)
0	22
1	30
2	38
3	46
4	54
5	62
6	70

In words: I started with 30 popsicle sticks and then I added/~~subtracted~~ 8 each time.
Find the mathematical rule:
Number of popsicle sticks = 8 x term number + 22
Tidy it up: $P = 8n + 22$
Word rule: The number of popsicle sticks is calculated by multiplying the term number by 8 and then adding/~~subtracting~~ 22.
Use the rule: $P = 8 \times 20 + 22$
$= 182$

5

Term number (n)	Number of dots (D)
0	40
1	36
2	32
3	28
4	24
5	20

In words: I started with 36 dots and then I ~~added~~/subtracted 4 each time.
Find the mathematical rule:
Number of dots = –4 x term number + 40
Tidy it up: $D = -4n + 40$
Word rule: The number of dots is calculated by multiplying the term number by –4 and then adding/~~subtracting~~ 40.
Use the rule: $D = -4 \times 8 + 40$
$= 8$

ISBN: 9780170451468

Finding the rule from a table (pp. 19–21)

1

Term number (n)	Value of term (T)
0	19
1	21
2	23
3	25
4	27

+ 2

Rule: $T = 2n + 19$

2

Term number (n)	0	1	2	3	4	5
Value of term (T)	15	19	23	27	31	35

+ 4

Rule: $T = 4n + 15$

3

Term number (n)	0	1	2	3	4	5
Value of term (T)	0	7	14	21	28	35

Rule: $T = 7n + 0$ or $T = 7n$

4

Term number (n)	Value of term (T)
0	12
1	24
2	36
3	48
4	60

Rule: $T = 12n + 12$

5

Term number (n)	0	1	2	3	4	5
Value of term (T)	54	52	50	48	46	44

Rule: $T = -2n + 54$

6

Term number (n)	Value of term (T)
0	82
1	79
2	76
3	76
4	70

Rule: $T = -3n + 82$

7

Term number (n)	1	2	3	4	5
Value of term (T)	–11	–4	3	10	17

Rule: $T = 7n - 18$

8

Term number (n)	1	2	3	4	5
Value of term (T)	101	98	95	92	89

Rule: $T = -3n + 104$

9

Term number (n)	Value of term (T)
1	27
2	18
3	9
4	0

Rule: $T = -9n + 36$

10

Term number (n)	1	2	3	4	5
Value of term (T)	–13	–15	–17	–19	–21

Rule: $T = -2n - 11$

Finding the rule from a list (pp. 22–23)

1 $T = 3n + 4$ **2** $T = -2n + 21$

3 $T = 5n - 4$ **4** $T = -4n + 28$

5 $T = -11n + 111$ **6** $T = 15n - 15$

7 $T = 4n - 23$ **8** $T = 6n - 9$

9 8, 11, 14, 17, …

–3 +3 +3 +3

I started with 8 and then I added/~~subtracted~~ 3 each time.

Term zero will be 5.

Rule: $T = 3n + 5$

10 I started with 1 and then I added/~~subtracted~~ 3 each time.
Term zero will be –2.
Rule: $T = 3n - 2$

11 I started with 23 and then I ~~added~~/subtracted 4 each time.
Term zero will be 27.
Rule: $T = -4n + 27$

12 I started with –12 and then I added/~~subtracted~~ 11 each time.
Term zero will be –23.
Rule: $T = 11n - 23$

13 I started with 126 and then I ~~added~~/subtracted 39 each time.
Term zero will be 165.
Rule: $T = -39n + 165$

14 Rule: $T = 45n - 62$

15 Rule: $T = 18n - 82$

16 Rule: $T = 99n - 198$

17 Rule: $T = 3.75n - 3.25$

Finding a term from a rule (p. 24)

1 $T = 335$
2 $T = 177$
3 $T = 67$
4 $T = -576$
5 $T = 126$
6 $T = -117$
7 $T = -1330$
8 $T = 393$
9 $T = -15$
10 $T = -9.25$

Finding the sequence from the rule (pp. 25–26)

1

Term number (n)	Calculations	Value of term (T)
1	4 x 1 + 7	11
2		15
3		19
4		23
5	4 x 5 + 7	27

Sequence: 11, 15, 19, 23, 27, …

2

Term number (n)	1	2	3	4	5
Calculations	–2 x 1 – 8				–2 x 5 – 8
Value of term (T)	–10	–12	–14	–16	–18

Sequence: –10, –12, –14, –16, –18, …

3

Term number (n)	Calculations	Value of term (T)
1	8 x 1 – 11	–3
2		5
3		13
4		21
5	8 x 5 – 11	29

Sequence: –3, 5, 13, 21, 29, …

4 Sequence: 15, 9, 3, –3, –9, …
5 Sequence: 0, 1, 2, 3, 4, …
6 Sequence: 1.5, 1, 0.5, 0, –0.5 , …
7 Sequence: –8, 7, 22, 37, 52 , …
8 Sequence: 83, 77, 71, 65, 59, …

Applications (pp. 27–28)

1 a 75, 100, 125, …
$T = 25n + 50$
b Amount = $1350

2 a 39, 53, 67, …
$T = 14n + 25$
b $179

3 a 493, 486, 479, …
$T = 500 - 7n$
b 262 staples

4 a 34, 30, 26, …
$T = -4n + 38$
b $6

5 a 153, 144, 135, …
$T = -9n + 162$
b 63 dresses

6 a 595, 630, 665, …
$T = 35n + 560$
b $1155

7 a 517, 469, 421, …
$T = -48n + 565$
b $181
c If there's no money left:
$0 = -48n + 565$
$n = \frac{565}{48}$
$n = 11.77$, so 11 mallets

ISBN: 9780170451468

Graphs (pp. 29–67)

Coordinate revision (pp. 29–30)

A (-4, 9) **B** (–8, –4)
C (8, 0) **D** (0, 0)
E (–7, 0) **F** (0, –7)
G (–3, –1) **H** (6, 7)

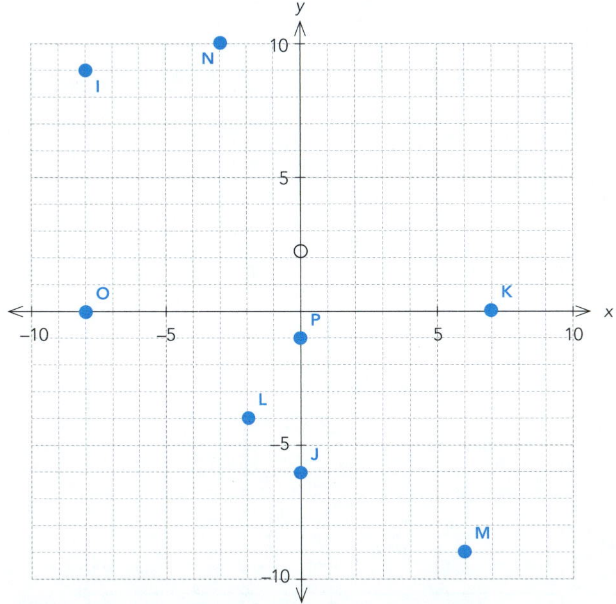

Plotting discrete patterns (pp. 31–34)

1

Term number (x)	Value of the term (y)
1	6
2	**9**
3	**12**
4	**15**
5	**18**

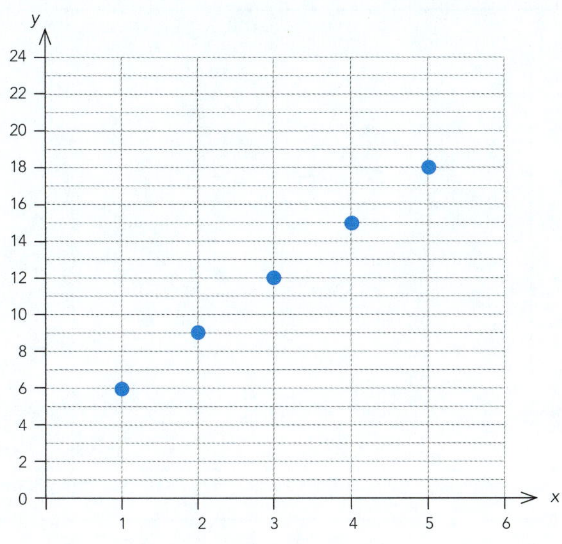

$y =$ **3** x term number + **3**

Tidy it up: $y = 3x + 3$

2

Term number (x)	Value of the term (y)
1	**18**
2	**15**
3	**12**
4	**9**
5	**6**

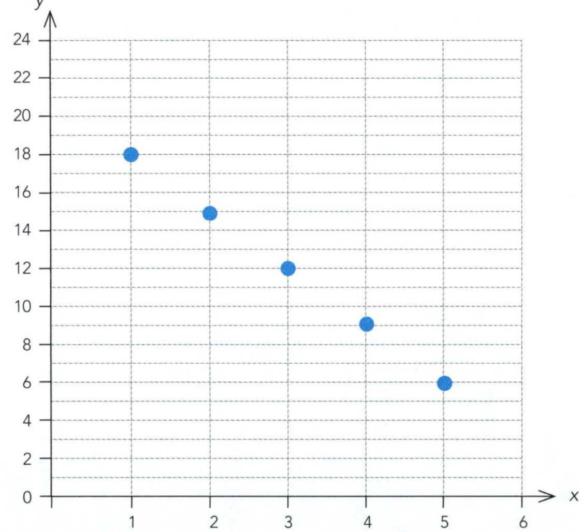

$y =$ **–3** x term number + **21**

Tidy it up: $y = -3x + 21$

3

Term number (x)	Value of the term (y)
1	6
2	11
3	**16**
4	**21**
5	**26**

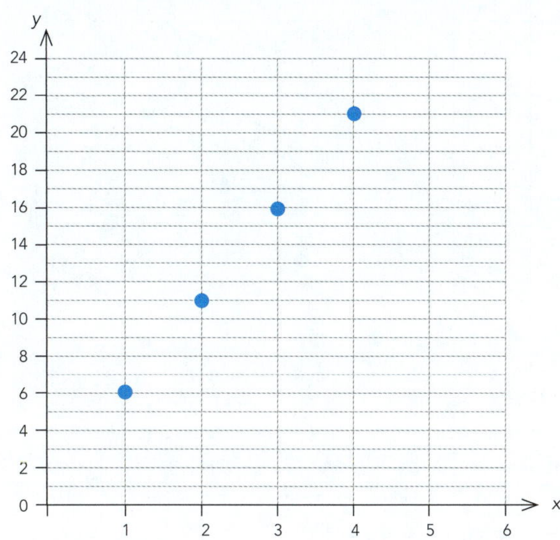

Equation: $y = 5x + 1$

ISBN: 9780170451468

4

Term number (x)	Value of the term (y)
1	20
2	**17**
3	14
4	**11**
5	**8**

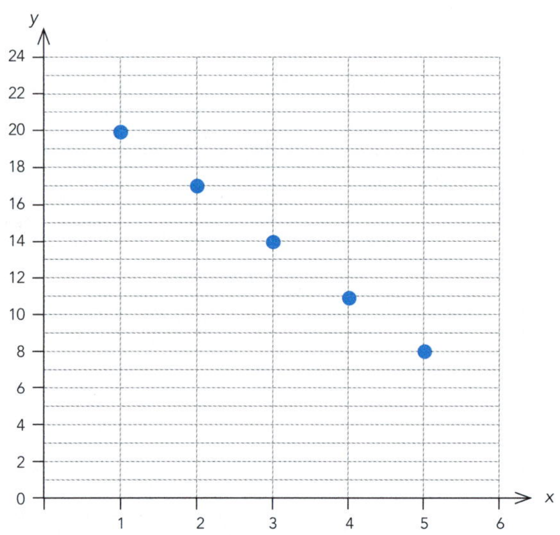

Equation: $y = -3x + 23$

5

Term number (x)	Value of the term (y)
1	3
2	**7**
3	**11**
4	**15**
5	19

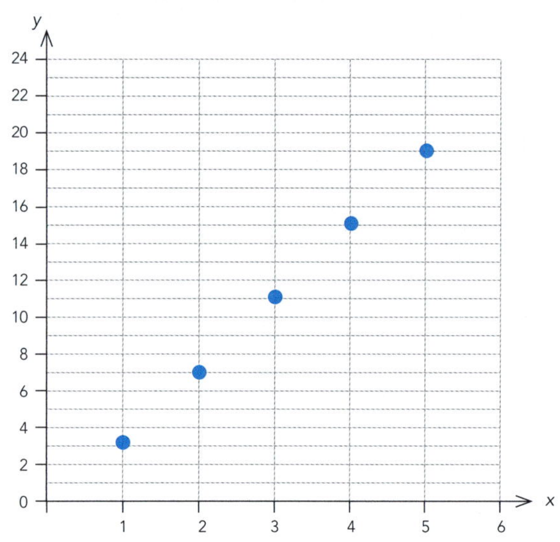

Equation: $y = 4x - 1$

Discrete applications (pp. 35–36)

1

Bags (x)	Money (y)
1	**$10**
2	**$14**
3	**$18**
4	**$22**
5	**$26**

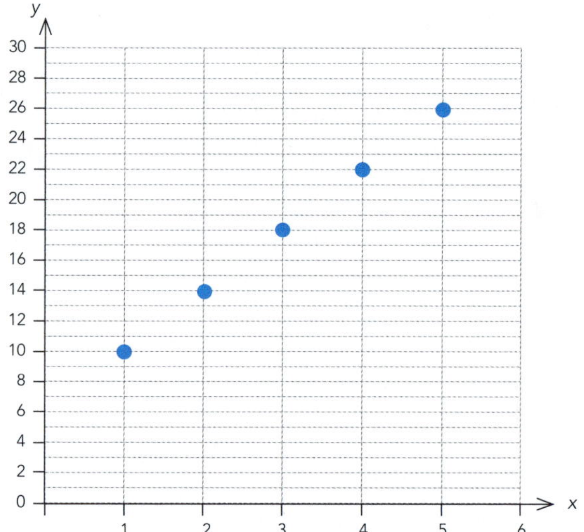

Equation: $y = 4x + 6$

$54

2

Weeks (x)	Amount owed (y)
1	**$120**
2	**$105**
3	**$90**
4	**$75**
5	**$60**

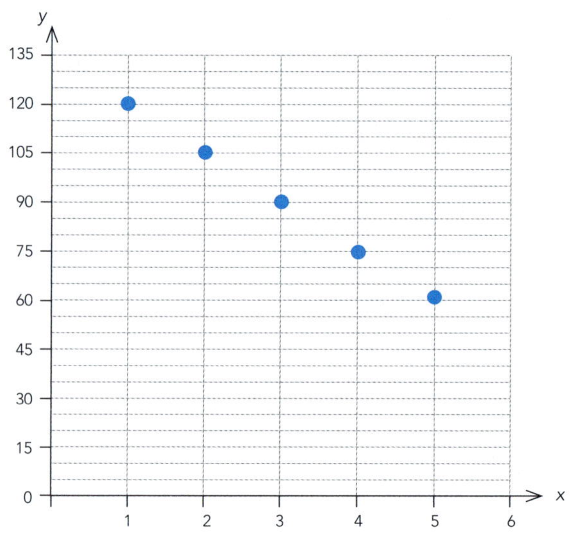

Equation: $y = -15x + 135$

$15

 ISBN: 9780170451468

3

Batch (x)	Cups of flour (y)
1	6
2	12
3	18
4	24
5	30

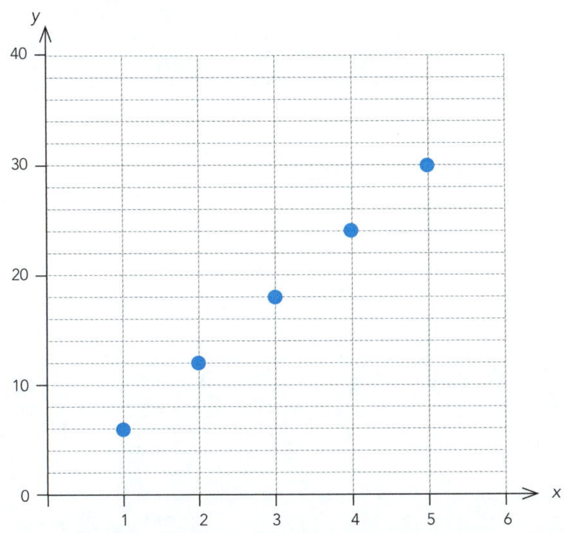

Equation: $y = 6x + 0$ or $y = 6x$

72

4

Days (x)	Tablets (y)
1	32
2	30
3	28
4	26
5	24

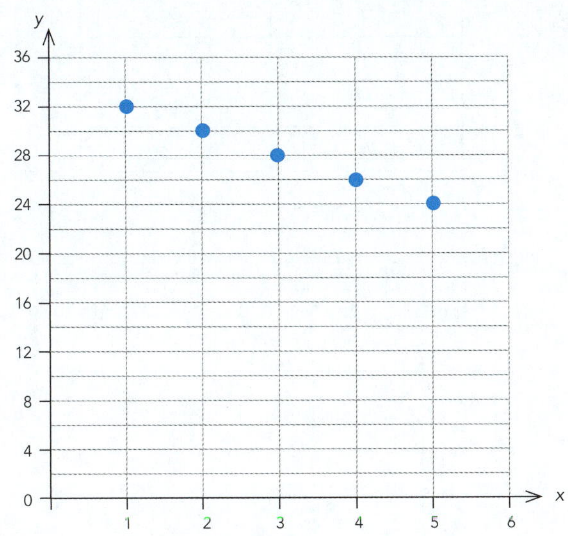

Equation: $y = -2x + 34$

8 tablets

Plotting continuous patterns (pp. 37–41)

1

x	Calculation	y	Coordinates
0	2 x 0 + 4	4	(0, 4)
1		6	(1, 6)
2		8	(2, 8)
3		10	(3, 10)
4		12	(4, 12)
5		14	(5, 14)
6	2 x 6 + 4	16	(6, 16)

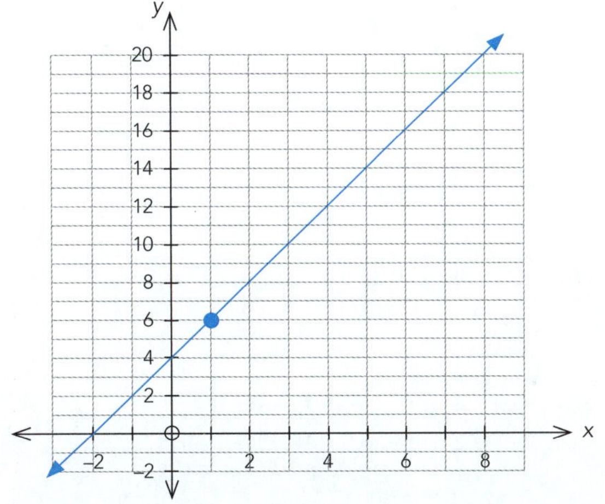

2

x	Calculation	y	Coordinates
0	–0 + 15	15	(0, 15)
1		14	(1, 14)
2		13	(2, 13)
3		12	(3, 12)
4		11	(4, 11)
5		10	(5, 10)
6	–6 + 15	9	(6, 9)

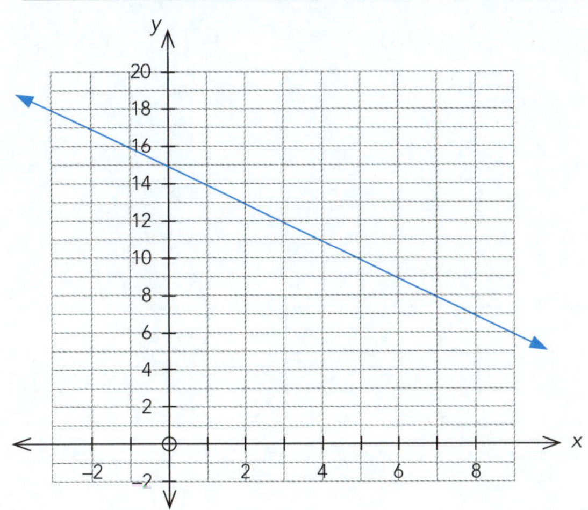

ISBN: 9780170451468

3

x	y
0	–5
1	–2
2	1
3	4
4	7
5	10
6	13

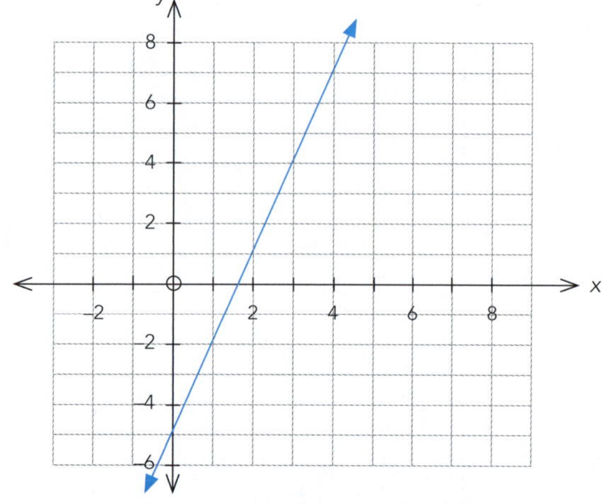

4

x	y
0	7
1	5
2	3
3	1
4	–1
5	–3
6	–5

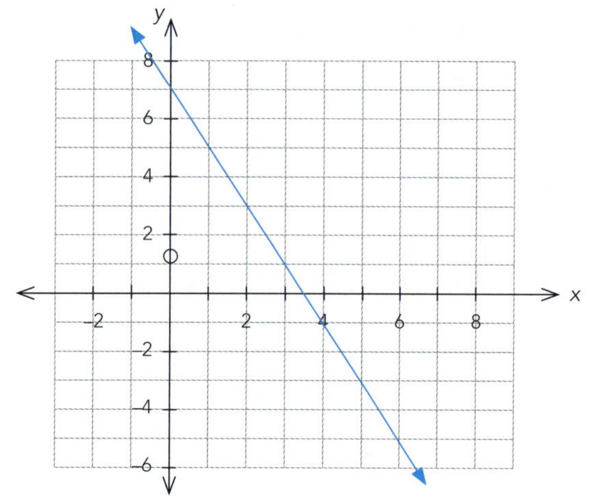

5

x	y
0	0
1	4
2	8
3	12
4	16
5	20
6	24

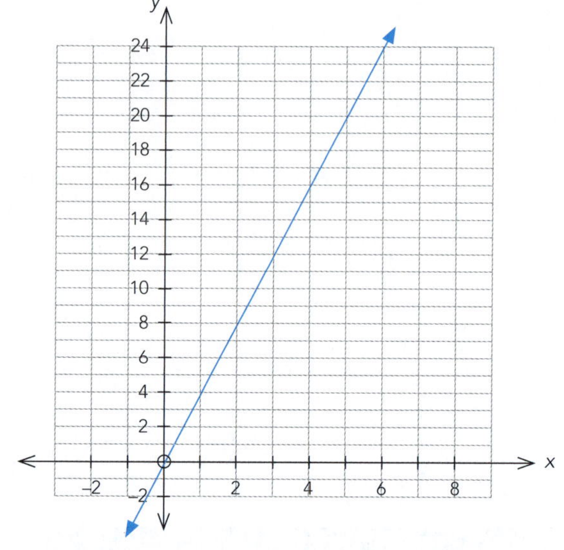

6

x	y
0	0
1	–2
2	–4
3	–6
4	–8
5	–10
6	–12

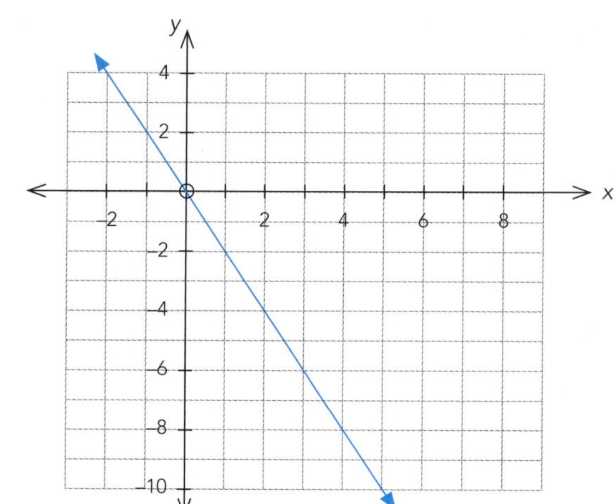

 ISBN: 9780170451468

7

x	y
0	–1
1	–2
2	–3
3	–4
4	–5
5	–6
6	–7

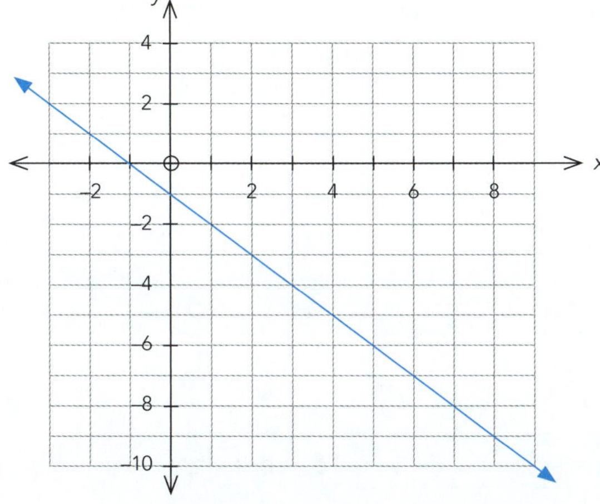

8

x	y
0	1
1	–2
2	–5
3	–8
4	–11
5	–14
6	–17

9

x	y
0	–3
1	–5
2	–7
3	–9
4	–11
5	–13
6	–15

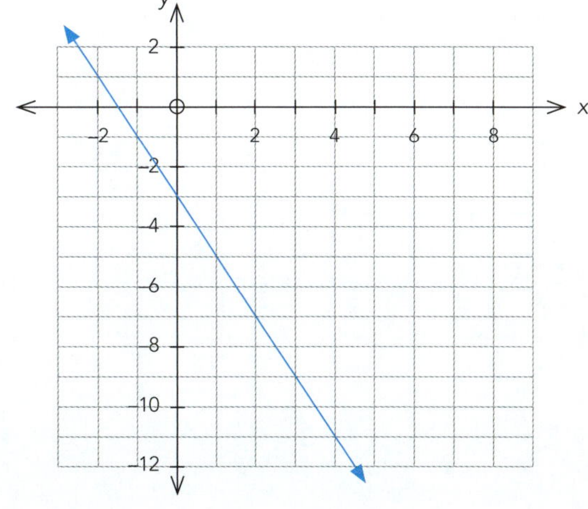

Finding linear equations from graphs (pp. 42–43)

1

x	y
0	–1
1	3
2	7
3	11
4	15

Equation:
$y = 4x - 1$

2

x	y
0	10
1	7
2	4
3	1
4	–2

Equation:
$y = -3x + 10$

3

x	y
0	3
1	4
2	5
3	6
4	7

Equation:
$y = 1x + 3$
or $y = x + 3$

4

x	y
0	9
1	7
2	5
3	3
4	1

Equation:
$y = -2x + 9$

Horizontal and vertical lines (pp. 44–46)

1 $y = -1$

2 $x = 2$

3 $y = \frac{1}{2}$

4 $x = -1.5$

5

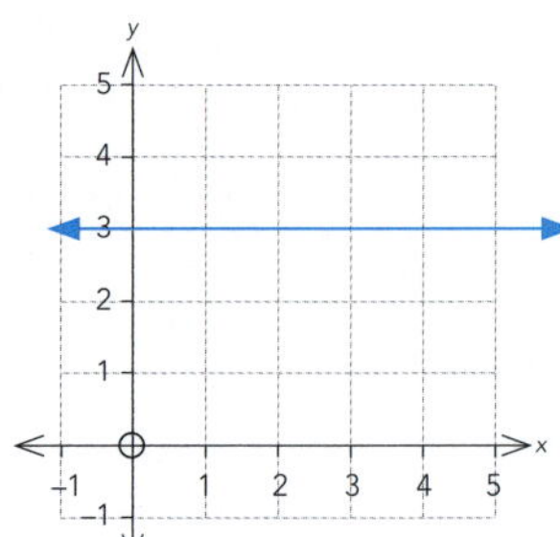

6

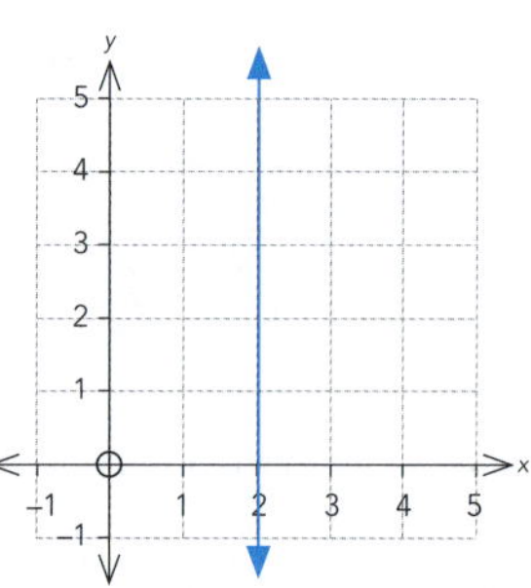
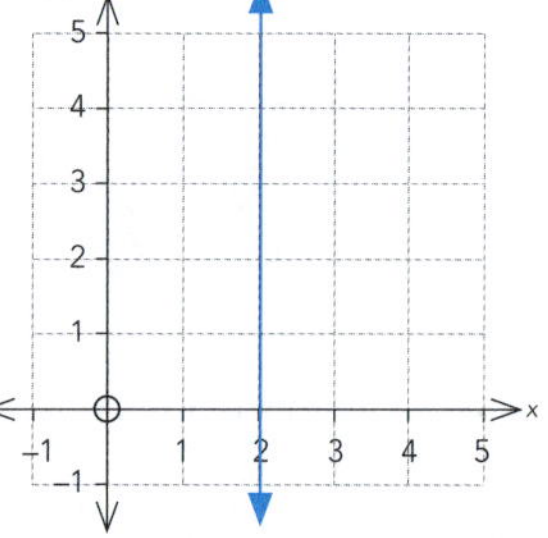

7

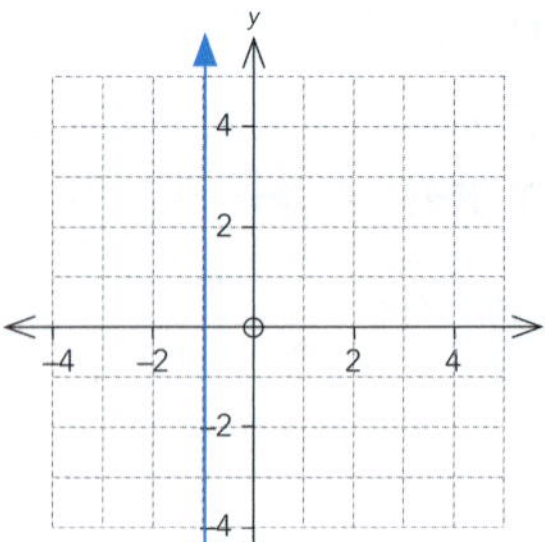

8

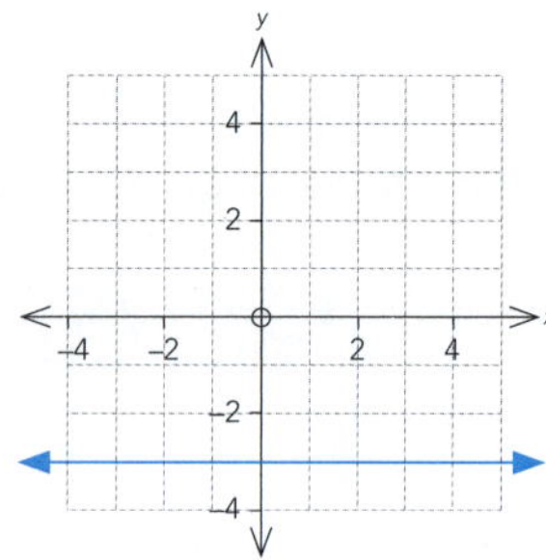

9

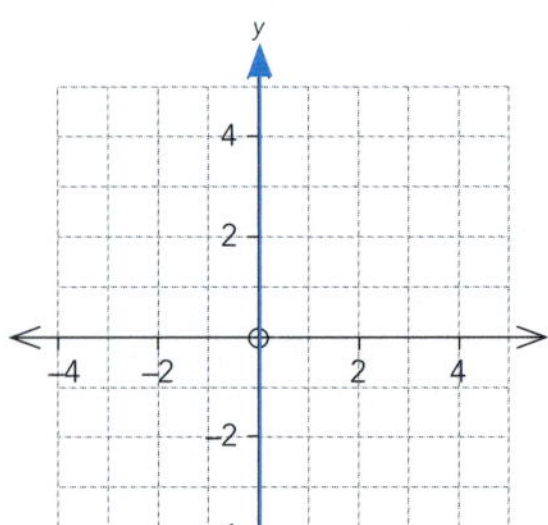

10

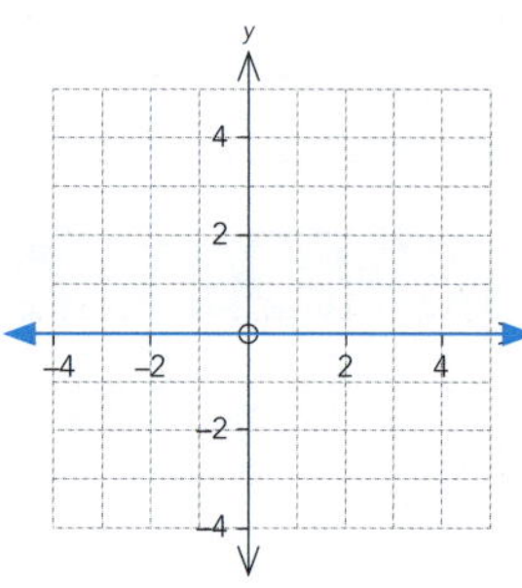

The gradient of the line (pp. 47–51)

1 $m = \frac{4}{3}$

2 $m = -\frac{2}{5}$

3 $m = -\frac{5}{2}$

4 $m = \frac{3}{1}$

5 $m = -\frac{3}{5}$

6 $m = \frac{1}{7}$

7 $m = \frac{4}{2}$

8 $m = -\frac{4}{4}$ or $-\frac{2}{2}$ or -1

9

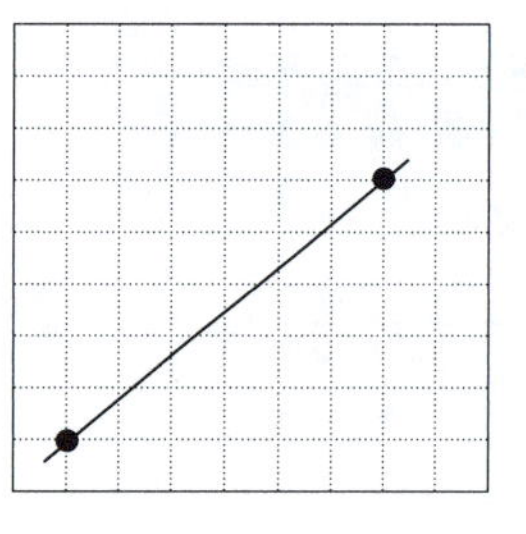

10

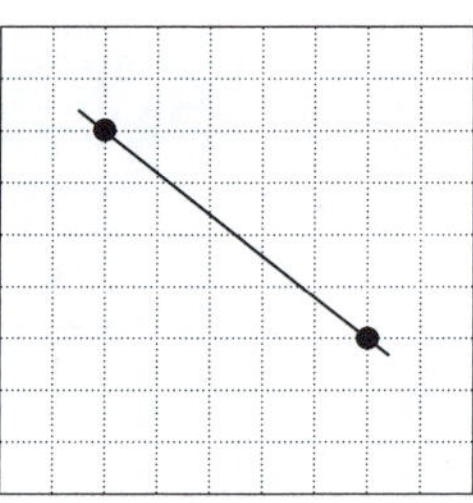

11

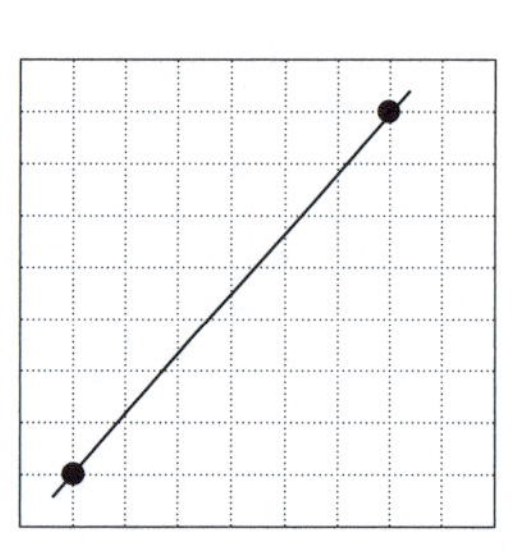

12

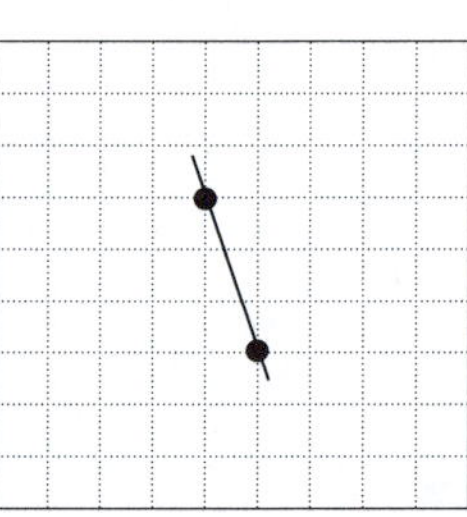

13

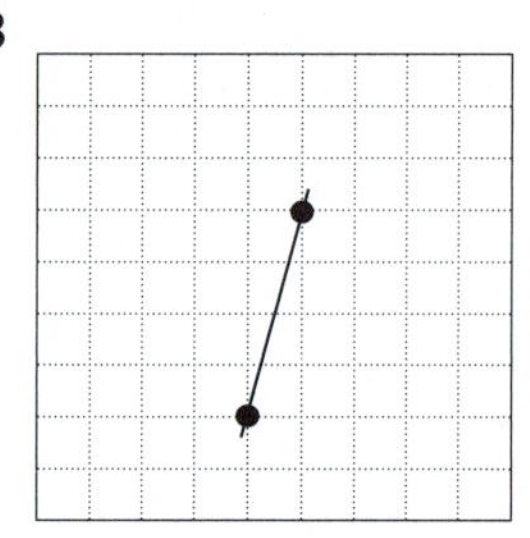

14

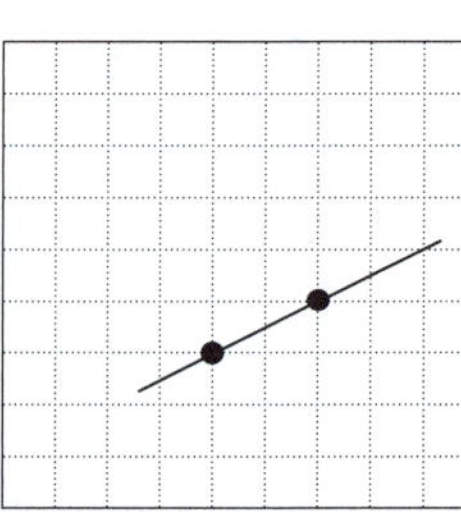

15 $m = \frac{4}{3}$

16 $m = -\frac{4}{1}$

17 $m = \frac{7}{2}$

18 $m = -\frac{2}{3}$

19 $m = \frac{3}{8}$

20 $m = -\frac{9}{5}$

Drawing straight lines using the gradient and y intercept (pp. 52–56)

1 $y = \frac{3}{4}x - 2$

2 $y = \frac{4}{3}x - 1$

3 $y = -\frac{1}{3}x + 4$

4 $y = -\frac{5}{2}x + 4$

5 $y = -3x - 1$

6 $y = \frac{9}{2}x - 2$

7

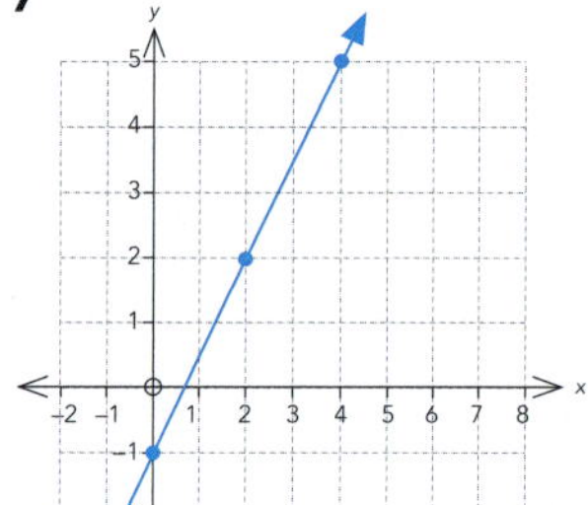

8

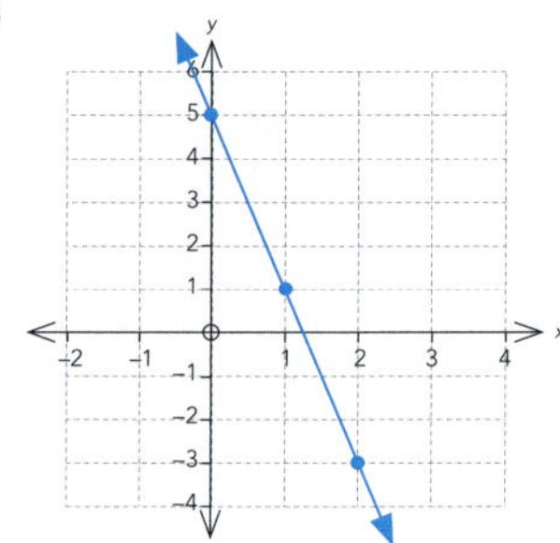

ISBN: 9780170451468

9

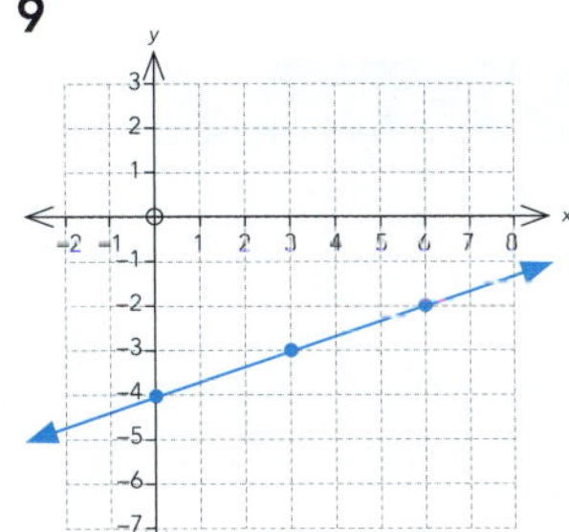

10

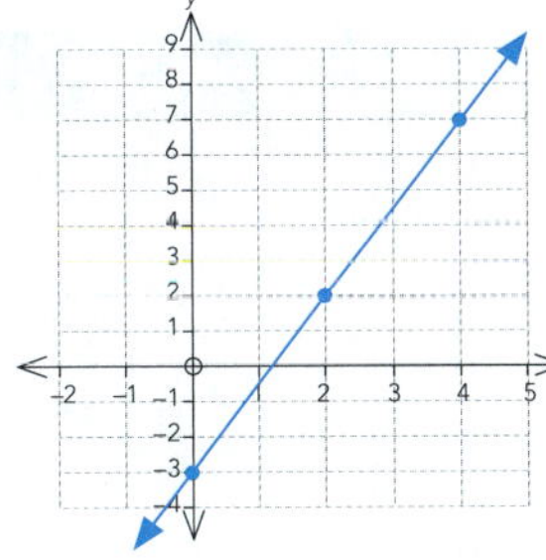

11

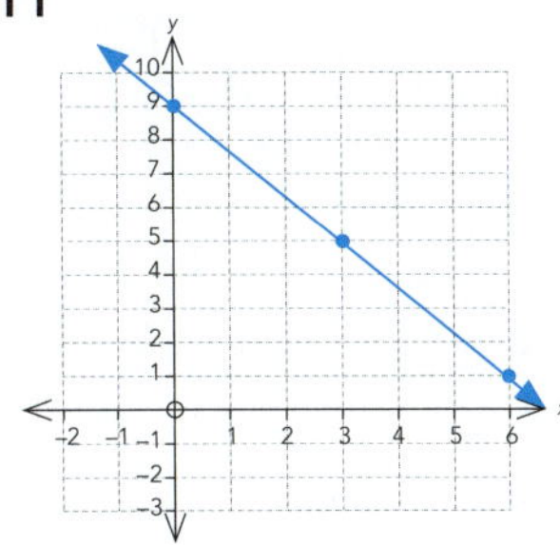

12

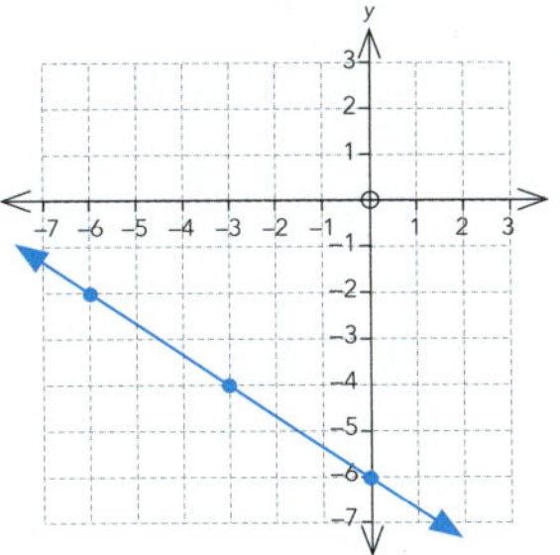

13

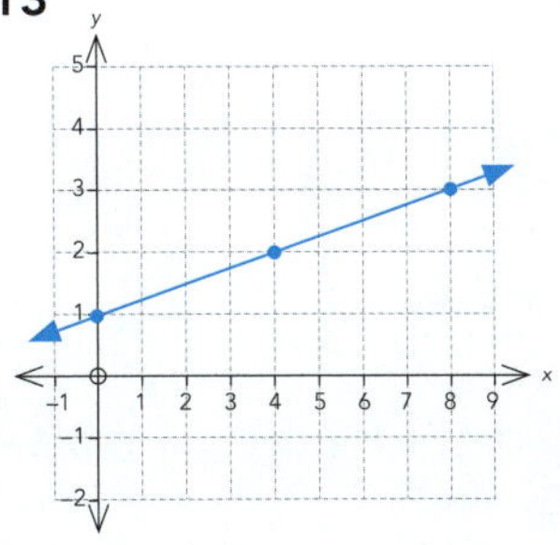

14

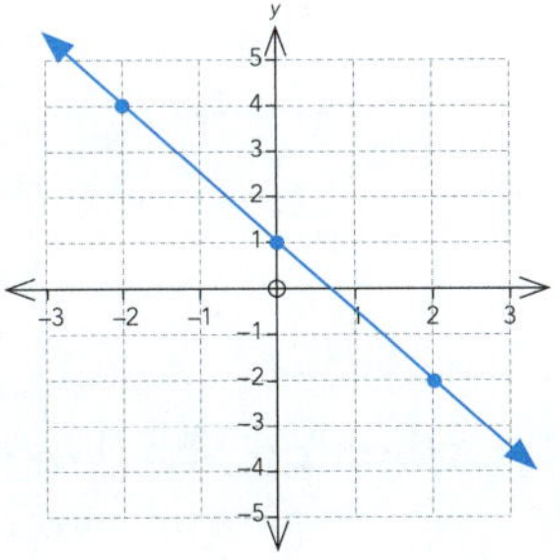

15

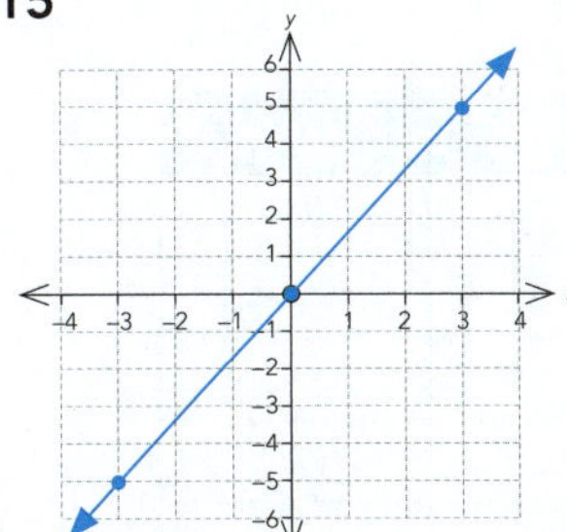

16

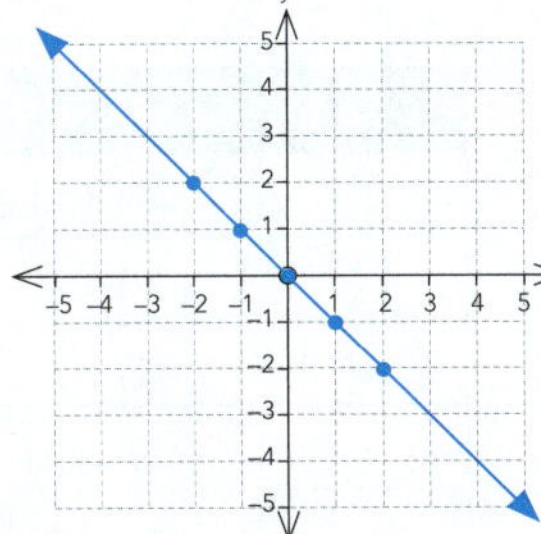

17

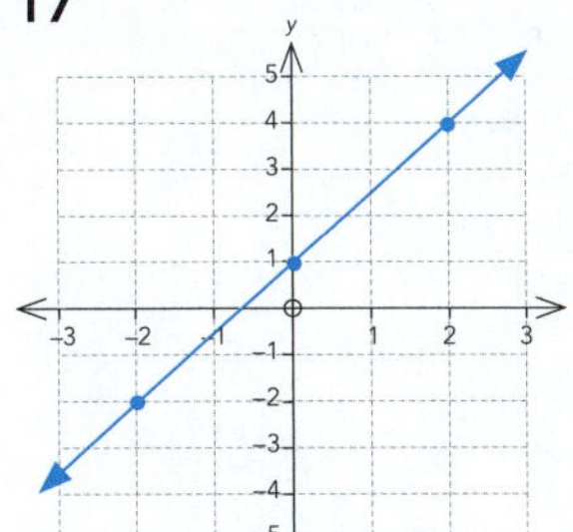

18

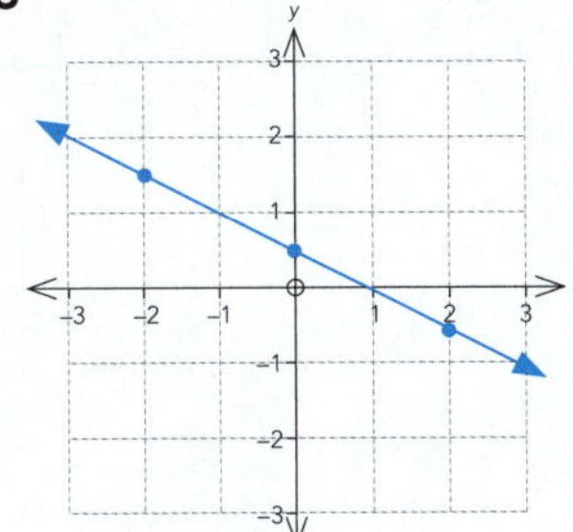

Writing equations from a graph (pp. 57–59)

1 $y = \frac{3}{2}x - 2$

2 $y = -\frac{6}{5}x + 7$

3 $y = \frac{1}{2}x + 3$

4 $y = \frac{2}{1}x$ or $y = 2x$

5 $y = -\frac{1}{6}x + 3$

6 $y = -\frac{2}{5}x - 5$

7 $y = -4x$

8 $y = \frac{2}{5}x + 1$

9 $y = -\frac{2}{3}x + 6$

10 $y = 3x - 1$

11 $y = -\frac{1}{4}x + 9$

12 $y = \frac{1}{2}x + \frac{1}{2}$

Applications (pp. 60–65)

1 **a**

Time (h) (x)	Distance to the finish line (km) (y)
0	42
1	34
2	26
3	18
4	10
5	2

b Get your teacher or a neighbour to check this.
c It means that he ran at 8 km per hour.
d It means that at the start, he had 42 km to run.
e $y = -8x + 42$
f $y = -8 \times 5 + 42$
$y = 2$
So he is 2 km from the finish line after 5 hours.
g Any value between 5 h and 10 m and 5 h and 20 min.

2 **a** Get your teacher or a neighbour to check this.
b It means that 2.5 litres of water was leaking out each day.
c It means that there was 1000 L of water in the spa pool to start with.
d $y = -2.5x + 1000$
e $y = -2.5 \times 365 + 1000$
$y = 87.5$ L

3 **a** Points: (16, 6) and (8, –6)
$m = \frac{6 - (-6)}{16 - 8}$
$= 1.5$
b It means that the temperature increases by 1.5°C per hour.
c It means that the temperature in the freezer was –18°C when the power cut started.
d $y = 1.5x - 18$
e $y = 1.5 \times 16 - 18$
$y = 3$°C
f The line would become steeper.

4 **a** Points: (34, 181) and (26, 169)
$m = \frac{181 - 169}{34 - 26}$
$= \frac{12}{8}$
$= \frac{3}{2}$
b It means that for every 1 cm increase in shoe print, there is a 1.5 cm increase in height.

ISBN: 9780170451468

c $y = \frac{3}{2}x + 130$

d $y = \frac{3}{2}x + 130$

$= \frac{3}{2} \times 31 + 130$

$= 176.5$ cm

e Any value between 33.1 cm and 33.5 cm.

5 a $m = \frac{1}{20}$

b It means that for every increase in 20 g in mass of the chicken, you must cook it for one extra minute.

c $y = \frac{1}{20}x + 15$

d $y = \frac{1}{20} \times 1600 + 15$

$= 95$ min

e It means that the gradient decreases, which means it takes less time to cook the chicken.

Linear or not? (pp. 66–67)

1 Linear **2** Non-linear
3 Non -linear **4** Linear
5 Linear **6** Non-linear
7 Linear

Quadratic patterns (pp. 68–77)

Drawing parabolas (pp. 70–77)

1 In the form $y = x^2 \pm b$ (pp. 70–72)

1

x	$x^2 + 1$	y	Point
–3	$(-3)^2 + 1$	10	(–3, 10)
–2	$(-2)^2 + 1$	5	(–2, 5)
–1	$(-1)^2 + 1$	2	(–1, 2)
0	$(0)^2 + 1$	1	(0, 1)
1	$(1)^2 + 1$	2	(1, 2)
2	$(2)^2 + 1$	5	(2, 5)
3	$(3)^2 + 1$	10	(3, 10)

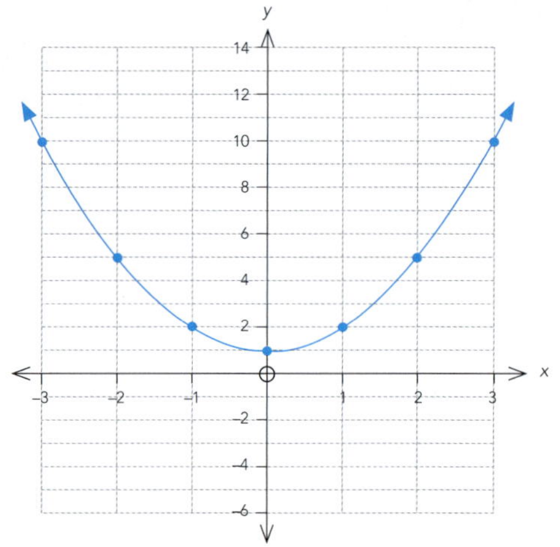

Compared with the graph of $y = x^2$, the parabola $y = x^2 + 1$ has moved up/~~down~~ 1.

2

x	$x^2 + 5$	y	Point
–3	$(-3)^2 + 5$	14	(–3, 14)
–2	$(-2)^2 + 5$	9	(–2, 9)
–1	$(-1)^2 + 5$	6	(–1, 6)
0	$(0)^2 + 5$	5	(0, 5)
1	$(1)^2 + 5$	6	(1, 6)
2	$(2)^2 + 5$	9	(2, 9)
3	$(3)^2 + 5$	14	(3, 14)

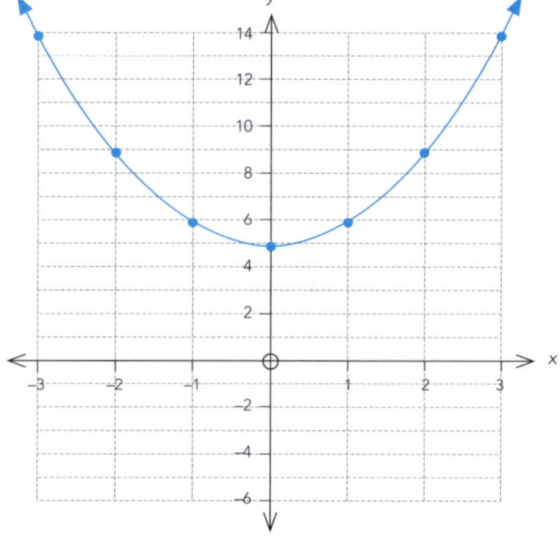

Compared with the graph of $y = x^2$, the parabola $y = x^2 + 5$ has moved up/~~down~~ 5.

3

x	$x^2 - 4$	y	Point
–3	$(-3)^2 - 4$	5	(–3, 5)
–2	$(-2)^2 - 4$	0	(–2, 0)
–1	$(-1)^2 - 4$	–3	(–1, –3)
0	$(0)^2 - 4$	–4	(0, –4)
1	$(1)^2 - 4$	–3	(1, –3)
2	$(2)^2 - 4$	0	(2, 0)
3	$(3)^2 - 4$	5	(3, 5)

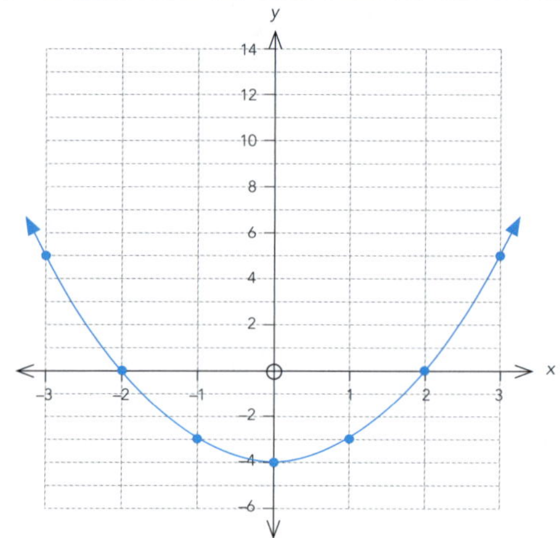

Compared with the graph of $y = x^2$, the parabola $y = x^2 - 4$ has moved ~~up~~/down 4.

 ISBN: 9780170451468

4

x	$x^2 - 6$	y	Point
–3	$(-3)^2 - 6$	3	(–3, 3)
–2	$(-2)^2 - 6$	–2	(–2, –2)
–1	$(-1)^2 - 6$	–5	(–1, –5)
0	$(0)^2 - 6$	6	(0, –6)
1	$(1)^2 - 6$	–5	(1, –5)
2	$(2)^2 - 6$	–2	(2, –2)
3	$(3)^2 - 6$	3	(3, 3)

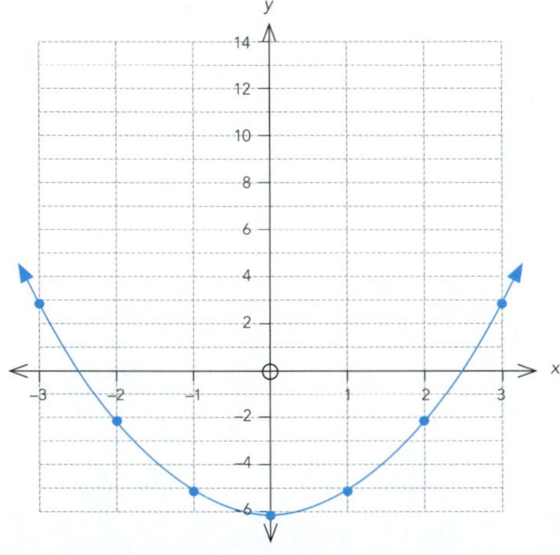

Compared with the graph of $y = x^2$, the parabola $y = x^2 - 6$ has moved ~~up~~/down 6.

2 In the form $y = (x \pm c)^2$ (pp. 73–76)

1

x	$(x + 1)^2$	y	Point
–3	$(-3 + 1)^2$	4	(–3, 4)
–2	$(-2 + 1)^2$	1	(–2, 1)
–1	$(-1 + 1)^2$	0	(–1, 0)
0	$(0 + 1)^2$	1	(0, 1)
1	$(1 + 1)^2$	4	(1, 4)
2	$(2 + 1)^2$	9	(2, 9)
3	$(3 + 1)^2$	16	(3, 16)
–4	$(-4 + 1)^2$	9	(–4, 9)

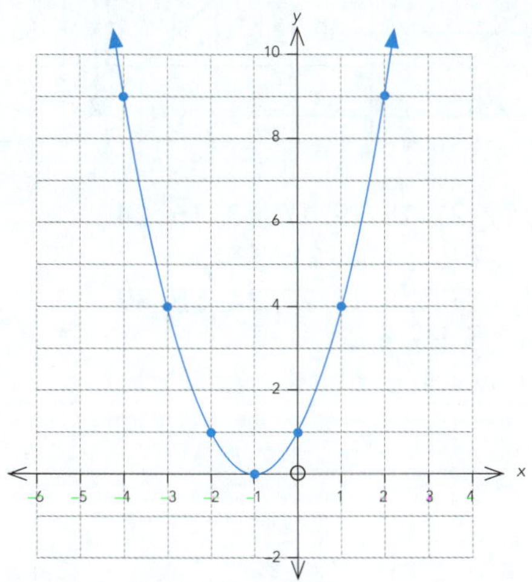

Compared with the graph of $y = x^2$, the parabola $y = (x + 1)^2$ has moved ~~right~~/left 1.

2

x	$(x + 4)^2$	y	Point
–4	$(-4 + 4)^2$	0	(–4, 0)
–3	$(-3 + 4)^2$	1	(–3, 1)
–2	$(-2 + 4)^2$	4	(–2, 4)
–1	$(-1 + 4)^2$	9	(–1, 9)
0	$(0 + 4)^2$	16	(0, 16)
–5	$(-5 + 4)^2$	1	(–5, 1)
–6	$(-6 + 4)^2$	4	(–6, 6)
–7	$(-7 + 4)^2$	9	(–7, 9)

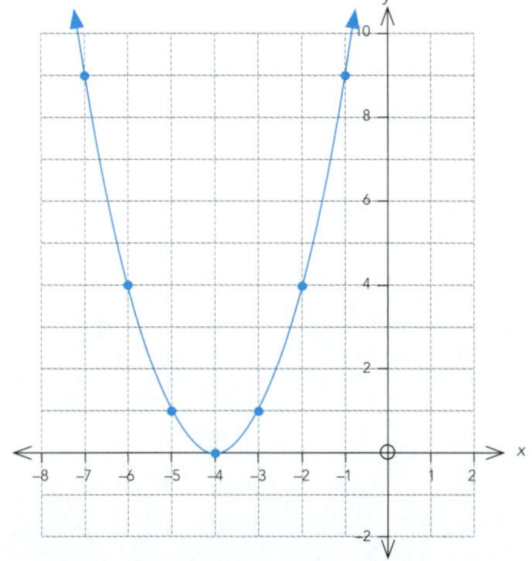

Compared with the graph of $y = x^2$, the parabola $y = (x + 4)^2$ has moved ~~right~~/left 4.

3 Some possible points:

x	$(x - 3)^2$	y	Point
3	$(3 - 3)^2$	0	(3, 0)
2	$(2 - 3)^2$	1	
1	$(1 - 3)^2$	4	
0	$(0 - 3)^2$	9	
4	$(4 - 3)^2$	1	
5	$(5 - 3)^2$	4	
6	$(6 - 3)^2$	9	

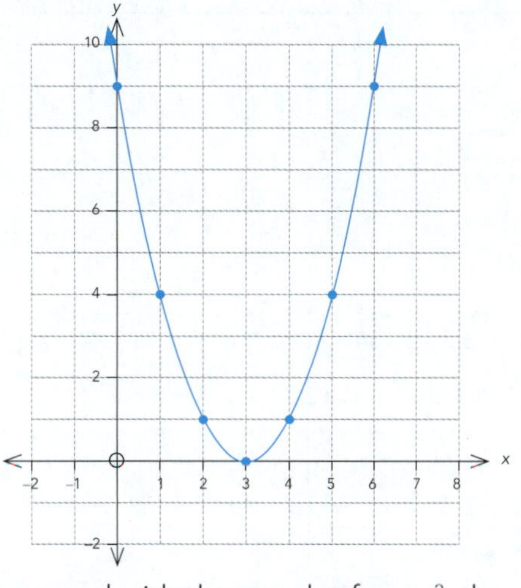

Compared with the graph of $y = x^2$, the parabola $y = (x - 3)^2$ has moved right/~~left~~ 3.

ISBN: 9780170451468

4 Some possible points:

x	$(x - 5)^2$	y	Point
5	$(5 - 5)^2$	0	
4	$(4 - 5)^2$	1	
3	$(3 - 5)^2$	4	
2	$(2 - 5)^2$	9	
6	$(6 - 5)^2$	1	
7	$(7 - 5)^2$	4	
8	$(8 - 5)^2$	9	

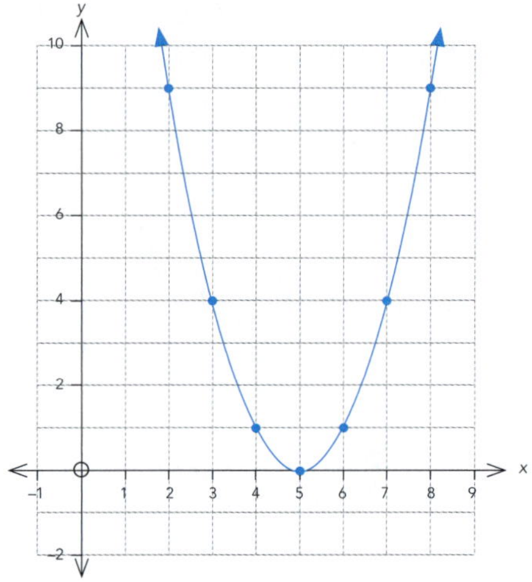

Compared with the graph of $y = x^2$, the parabola $y = (x - 5)^2$ has moved right/~~left~~ 5.

In the form

3 $y = -x^2$ (p. 77)

x	$-x^2$	y	Point
–3	$-(-3)^2$	–9	(–3, –9)
–2	$-(-2)^2$	–4	(–2, –4)
–1	$-(-1)^2$	–1	(–1, –1)
0	$-(0)^2$	0	(0, 0)
1	$-(1)^2$	–1	(1, –1)
2	$-(2)^2$	–4	(2, –4)
3	$-(3)^2$	–9	(3, –9)

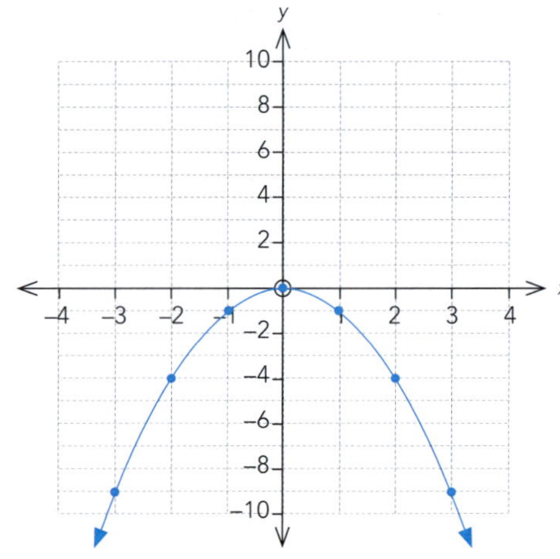

Compared with the graph of $y = x^2$, the parabola $y = -x^2$ is upside down.

Putting it together (p. 77)

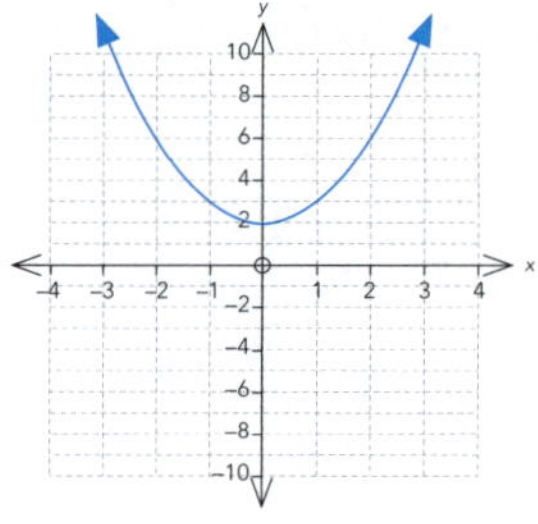

$y = x^2 + 2$

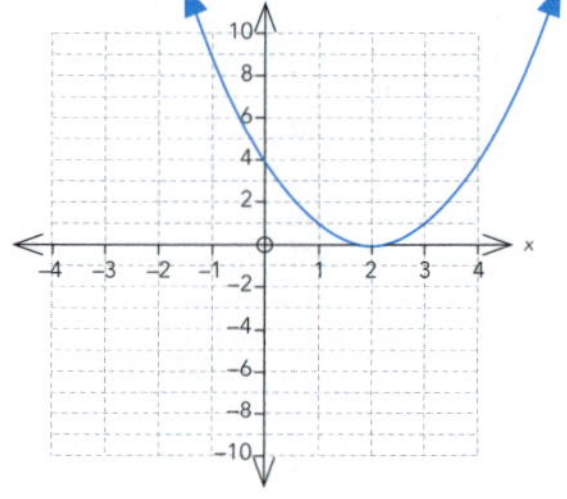

$y = (x - 2)^2$

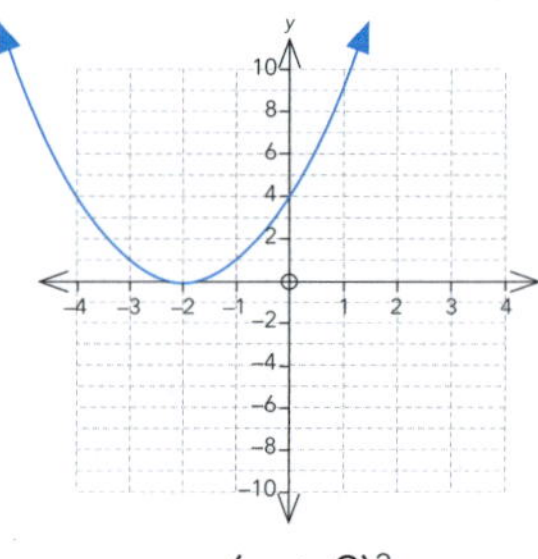

$y = (x + 2)^2$

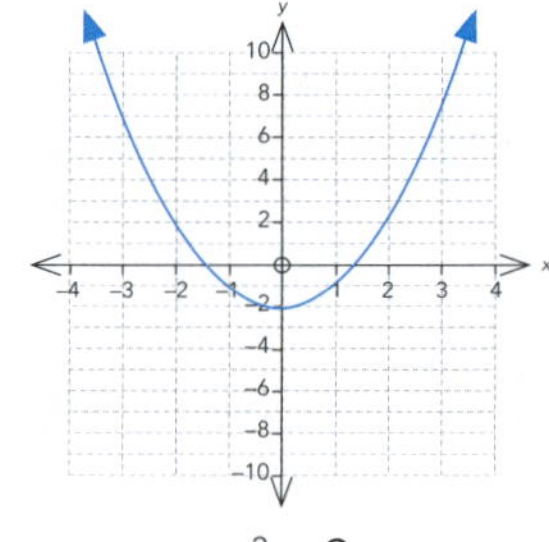

$y = x^2 - 2$

Challenge (pp. 78)

1

45	38	37	36	25	22	11	1	23	45
42	34	30	26	22	18	15	17	19	21
31	21	23	16	15	14	11	10	10	–3
21	17	14	11	8	5	1	4	1	4
41	23	15	7	–1	–9	–17	–10	–3	9
32	22	12	2	12	14	15	16	17	16
23	21	–11	–3	4	10	15	20	30	25
14	6	–18	–8	0	8	16	24	40	36
5	–1	–7	–13	–19	6	16	28	38	48
–3	–8	–20	–32	–44	–56	–66	32	34	60

2 **a** 20, 19, 21, 18, 22, 17, **23**, **16**, **24**

b 1, 8, 27, 64, **125**, **216**, **343**

c 1, 1, 2, 3, 5, 8, 13, 21, **34**, **55**, **89**

d o, t, t, f, f, **s**, **s**, **e**

e m, t, w, t, **f**, **s**, **s**

 ISBN: 9780170451468

Revision 1 (pp. 79–82)

1 **a**

Shape number (n)	Number of popsicle sticks (T)
0	5
1	12
2	19
3	26
4	33
5	40
6	47

b Number of popsicle sticks = **7** x shape number + **5**

T = **$7n$ + 5**

c T = **7 x 50 + 5**

= **355**

2

Term number (n)	0	1	2	3	4	5
Value of term (T)	101	98	95	92	89	86

Rule: $T = -3n + 101$

3 **a** Rule: $T = 5n + 8$

The 25th term = 133

b Rule: $T = -11n + 67$

The 19th term = –142

4

Term number (n)	1	2	3	4	5
Calculations	5 x 1 – 3				5 x 5 – 3
Value (T)	2	7	12	17	22

5 **a**

Days (x)	Mints (y)
0	28
1	25
2	22
3	19
4	16
5	13

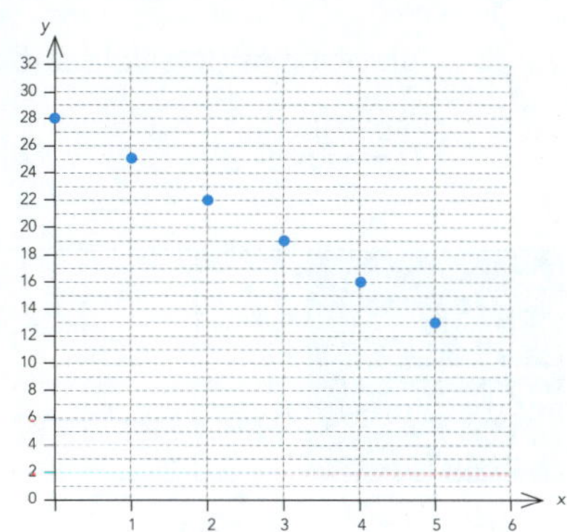

b $y = -3x + 28$

c 1 mint

6

x	y	Coordinates
0	–1	(0, –1)
1	3	(1, 3)
2	7	(2, 7)
3	11	(3, 11)
4	15	(4, 15)
5	19	(5, 19)
6	23	(6, 23)

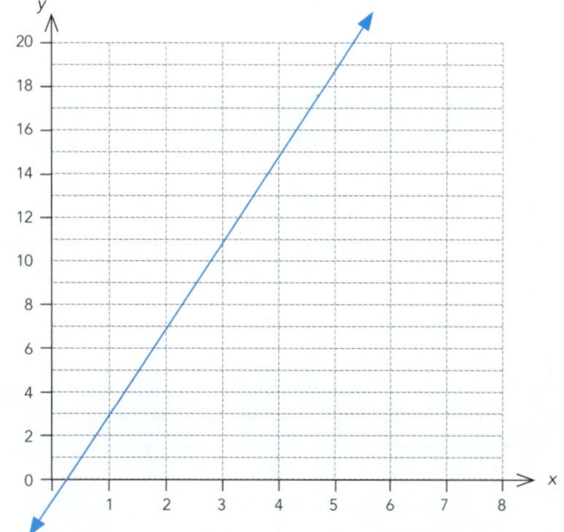

7

x	y
0	10
1	8
2	6
3	4
4	2

Equation: $y = -2x + 10$

8 **a** $m = \frac{3}{2}$

b

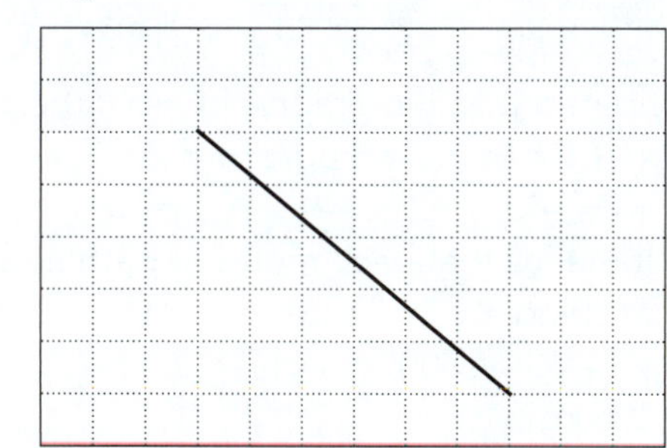

ISBN: 9780170451468

9 a

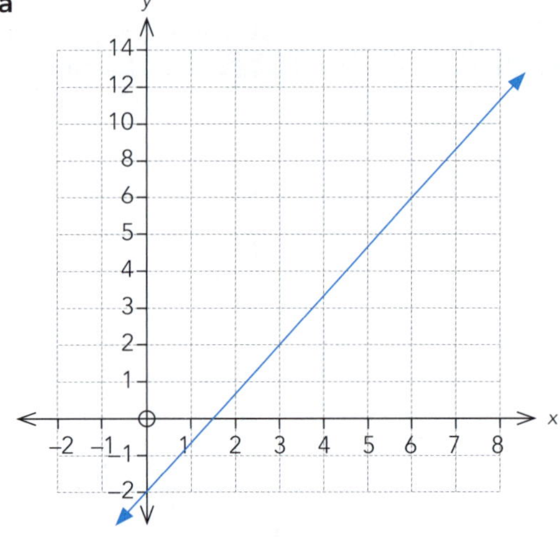

b $y = -3x + 13$

10

x	$x^2 + 2$	y	Point
–3	$(-3)^2 + 2$	11	(–3, 11)
–2	$(-2)^2 + 2$	6	(–2, 6)
–1	$(-1)^2 + 2$	3	(–1, 3)
0	$(0)^2 + 2$	2	(0, 2)
1	$(1)^2 + 2$	3	(1, 3)
2	$(2)^2 + 2$	6	(2, 6)
3	$(3)^2 + 2$	11	(3, 11)

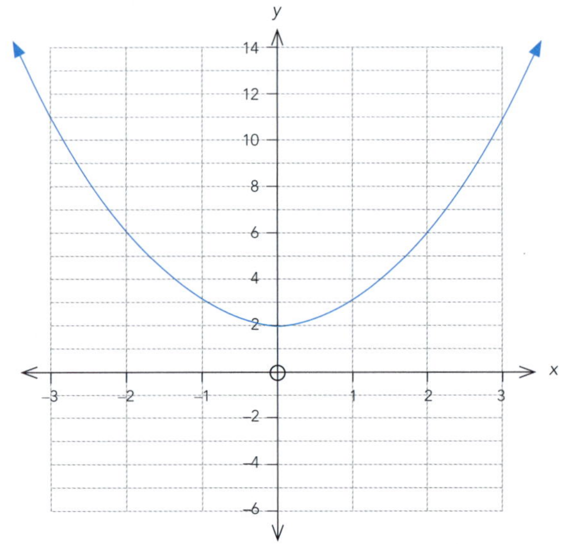

Compared with the graph of $y = x^2$, the parabola $y = x^2 + 2$ has moved up/~~down~~ 2.

11 a m = –45

b It means that each metre of path needs 45 kg of shingle.

c $y = -45x + 600$

d 195 kg

12 a

x	$(x-2)^2$	y	Point
–3	$(-3-2)^2$	25	(–3, 11)
–2	$(-2-2)^2$	16	(–2, 6)
–1	$(-1-2)^2$	9	(–1, 3)
0	$(0-2)^2$	4	(0, 2)
1	$(1-2)^2$	1	(1, 3)
2	$(2-2)^2$	0	(2, 6)
3	$(3-2)^2$	1	(3, 11)

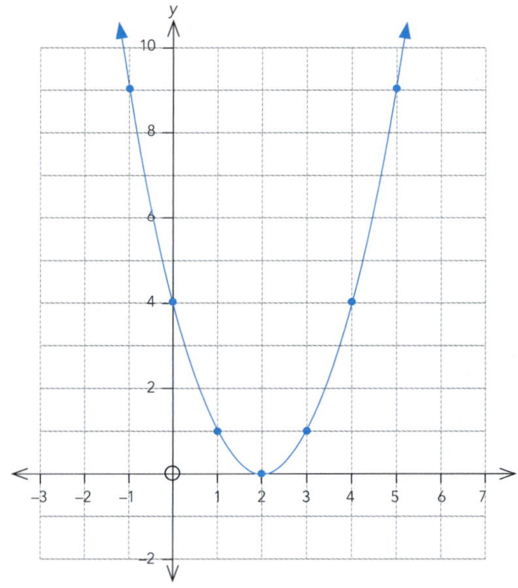

b Compared with the graph of $y = x^2$, the parabola $y = (x-2)^2$ has moved right/~~left~~ 2.

Revision 2 (pp. 83–86)

1 a

Shape number (*n*)	0	1	2	3	4
Number of popsicle sticks (*T*)	29	22	15	8	1

b Number of popsicle sticks = **7** x shape number + **29**

$T = -7n + 29$

2

Term number (*n*)	0	1	2	3	4	5
Value of term (*T*)	91	84	77	70	63	56

Rule: $T = -7n + 91$

  ISBN: 9780170451468

3 **a** Rule: $T = 8n - 10$

The 36th term = 278

b Rule: $T = -9n + 46$

The 15th term = -89

4

Term number (n)	1	2	3	4	5
Calculations	$-2 \times 1 + 6$				$-2 \times 5 + 6$
Value (T)	4	2	0	–2	–4

5 **a**

Rows (x)	Stitches (y)
0	4
1	6
2	8
3	10
4	12
5	14

b $y = 2n + 4$

c 20 stitches

6

x	y	Coordinates
0	**16**	**(0, 16)**
1	14	(1, 14)
2	12	(2, 12)
3	10	(3, 10)
4	8	(4, 8)
5	6	(5, 6)
6	4	(6, 4)

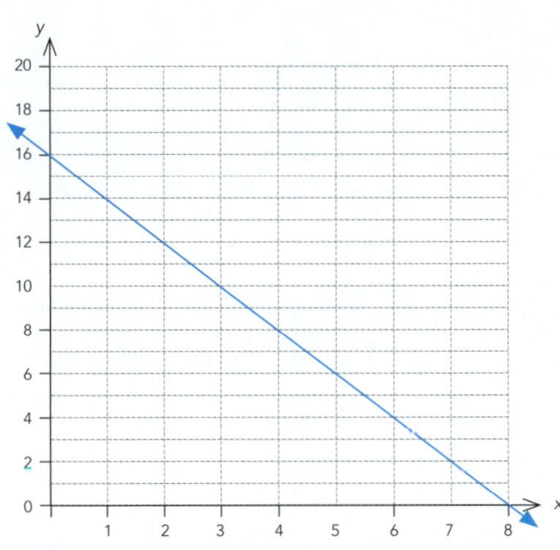

7

x	y
0	1
1	3
2	5
3	7
4	9

Equation: $y = 2x + 1$

8 **a** $m = -\frac{1}{4}$

b

9 **a**

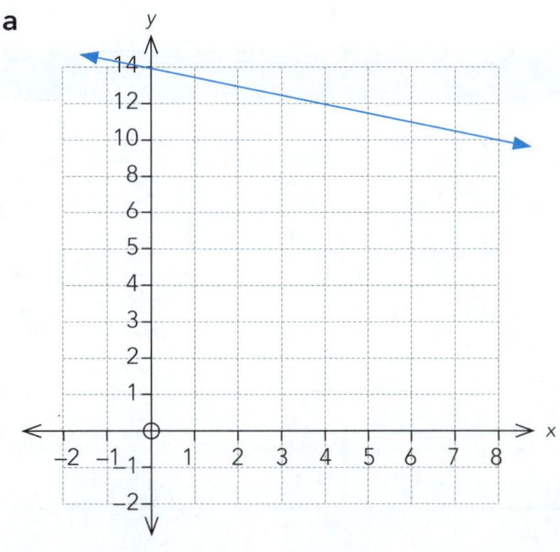

b $y = \frac{5}{2}x - 2$

10

x	$x^2 - 5$	y	Point
–3	$(-3)^2 - 5$	4	(–3, 4)
–2	$(-2)^2 - 5$	–1	(–2, –1)
–1	$(-1)^2 - 5$	–4	(–1, –4)
0	$(0)^2 - 5$	–5	(0, –5)
1	$(1)^2 - 5$	–4	(1, –4)
2	$(2)^2 - 5$	–1	(2, –1)
3	$(3)^2 - 5$	4	(3, 4)

ISBN: 9780170451468

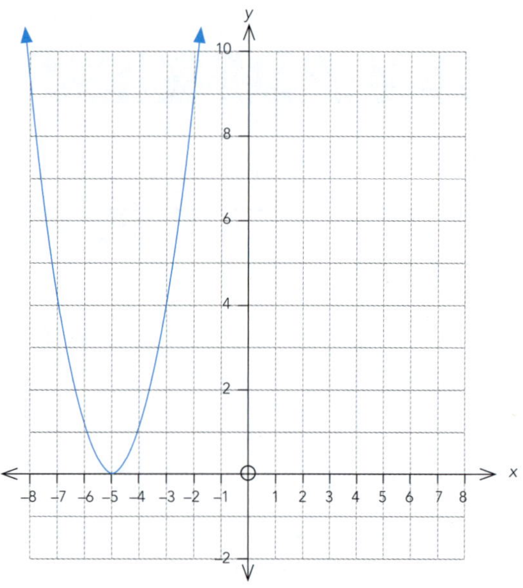

Compared with the graph of $y = x^2$, the parabola $y = x^2 - 5$ has moved up/~~down~~ 5.

11 **a** m = 2.5

b It means that each centimetre of the scarf needs 2.5 m of wool.

c $y = 2.5x$

d 300 m

12 **a**

x	$(x + 3)^2$	y	Point
–5	$(-5 + 3)^2$	4	(–5, 11)
–4	$(-4 + 3)^2$	1	(–2, 11)
–3	$(-3 + 3)^2$	0	(–3, 11)
–2	$(-2 + 3)^2$	1	(–2, 6)
–1	$(-1 + 3)^2$	4	(–1, 3)
0	$(0 + 3)^2$	9	(0, 2)
1	$(1 + 3)^2$	16	(1, 3)

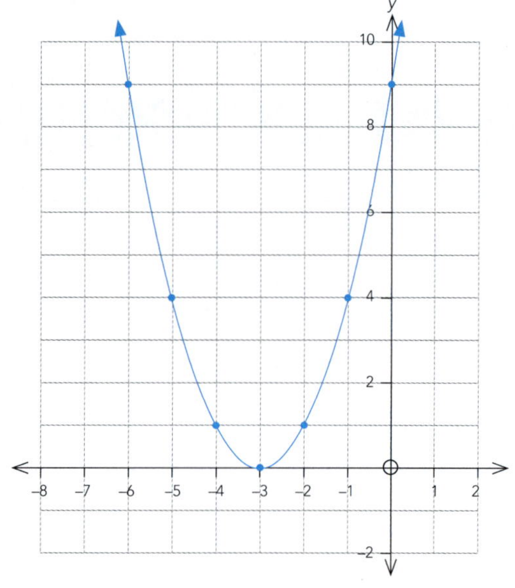

b Compared with the graph of $y = x^2$, the parabola $y = (x + 3)^2$ has moved ~~right~~/~~left~~ 3.

 ISBN: 9780170451468